Dave did a good job capturing the history and feel of Microsoft at a time when the wind blew in our hair and our sail was full. Admittedly, we made a lot of mistakes but in all fairness, we also got a lot of things right in those years and we never backed down from the challenges. We made enough mid-course corrections to bring the ship into port in good order.

Bob O'Rear, Microsoft Employee #7

Couldn't put it down! Fascinating... filled in many blanks about MSFT lore!! Awesome and Amazing insight for building business... from startup to Fortune 500...!!

Tim Waters, Nordstrom Corp, Former Digital Imaging Exec, 4th Generation Northern Californian, Guitar player...

Living in rural New Hampshire, the advances in technology meant that I could compete nationally with my direct sales company instead of ending up "on the farm". It is fascinating to read Jaworski's account of some of the events that have made my life today possible!

Lisa M. Wilber, one of only five Avon Platinum Executive Leaders in the United States, top 10 money earner for more than 2 decades — from Weare, New Hampshire

I loved Microsoft Secrets. The stories are fabulous. I enjoyed getting to know Dave better and to learn all that he experienced as an early member of Microsoft. And I especially appreciated the leadership and management lessons with questions. They made me think deeply about how I can be a stronger leader.

Patricia Shea, CEO, YWCA of Nas

Dave Jaworski's 'Microsoft Secrets' is fascinating and telling. The first-hand recollection true stories of how computer technology was born and grew into something world-changing which we can learn from today. Risk, reward, international politics, sex-drugs-rock'n roll, it's all in these pages, with great lessons any leader would like to see.

Joseph J. Simon, LUTCF, Business Development, Aon Risk Solutions

Dave Jaworski succeeded in his tenure during the early days of Microsoft, through myriad challenges, because of his faith-centered and integrity-driven personal "operating-system". This incredible page-burner of a book outlines more than just "secrets" of Gates, MS, and co. it tells the story of a man determined not to compromise his integrity for more stock options, and outlines the price he paid for openly sharing his values in the workplace. Bravo DJ. I highly recommend this book and the lessons learned he shares after each chapter for those facing similar challenges today.

David Pack, Grammy winning Artist-Producer, Former Singer-Co-Founder Prog-Rock Band Ambrosia, Rick Warren/Saddleback Church Artist for PEACE

I've had the pleasure of knowing Dave for over ten years now, and the privilege of working with him and witnessing his leadership and demeanor firsthand. He is someone who walks his talk, operates with integrity and treats his people with the utmost care and respect. He's always seeking a win for everyone involved in a transaction or experience — personally and professionally. I was an Apple exec, and many of the attributes that Dave describes about working in the Microsoft culture were similar to how we operated in the Apple culture — including the emphasis on hiring A players, workaholism, a dedication to be the best, and a commitment to delivering the best products we could for our customers. Both of us had the opportunity to 'ride a rocket ship' at two companies who literally changed the world. Dave shares many great insights & core principles from his experiences in the book that leaders in any business that aspires to be successful should take to heart and adopt to their own benefit and results.

Kelli Richards, President & CEO, The All Access Group LLC

Years ago, I learned to use Lotus 1-2-3 on an IBM PC and always wondered how Excel came out of nowhere to be the king of spreadsheets – Microsoft Secrets gave me the answer.

Steve Minucci, Regional President, Accelerent

Simply compelling! Dave shares how Microsoft put PCs in all our lives — to change us forever.

Blair Bryant, CEO, Arcadia Pacific providing leadership, advisory, and strategy consulting to Apple, Google, Microsoft, Porsche Design, McKinsey; Board of Directors, ThunderMaps

Dave does more than share his personal perspective as an insider at Microsoft; he transports you back to an era of the computer industry full of excitement and opportunity when the lines defining corporate culture were still evolving and offers practical insights helpful for today.

Eric Busby, President at CEadvisory, Inc.

The insights you will glean from Microsoft Secrets provide truly valuable lessons on leadership and strategic business principles for any industry, not just the Tech field. The author's passion for helping people and organizations achieve their full potential shines through this intriguing story by a brilliant insider. Dave provides powerful and easy-to-implement strategies in the "Take Action!" sections that can help us all build better companies and better lives.

Sandra Garest, President & Lifestyle Coach – Lifebook, Chicago (mylifebook.com)

This is an indispensable read for anyone who wishes to start or build a business with compassion and integrity.

Paul Roy MD FRCPC, Psychiatrist, Ottawa, Canada

Many today can't imagine life prior to the technology revolution. Dave Jaworski found himself in the epicenter of the revolution. Finishing college with a degree that he never understood was a ticket to make history, he knew Steve and Bill when they still required a last name to be recognized. This book is captivating because is it puts you in the center of this transformation. If you are too young to have experienced this while it was happening, this book will be a time machine giving you an appreciation for the amazing journey from where it all began.

Steve Johnson, President of 2xGlobal, Consultant, Speaker and author of TACTIC Evaluation (www.TACTICevaluation.com) and Live Happily Ever After (www.LiveHappilyEverAfter.com)

I like to describe my work with patients as "low tech and high touch". But truth be told, as a chiropractor, I get to work with the most impressive technology known to man – the human nervous system. Although our method of care is simple and elegant - and really only requires the use of our hands – technology has changed the delivery, efficiency and quality of care more than any other factor. Electronic patient record keeping, diagnostic tools, information accessibility and digital communications have elevated the game beyond measure. Dave Jaworski's new book, Microsoft Secrets, gives us all a fascinating peak behind the curtain at the story that shaped the technology that made this all possible.

Stephen Franson, DC, Founder, The Remarkable Practice

It was a privilege to work with Dave at Microsoft. Dave has the uncanny ability to know when it's time for a company to change course, how to get the boat heading in the right direction and, more importantly, get everyone in the boat rowing together. This book not only provides insights on lesser known but significant changemakers at Microsoft, it's a chance for others to be coached by one of the most astute mentors in the industry. Thank you Dave for writing this book and for the years of wise counsel you've generously extended to me and so many others.

Lisa Scattaregia, Deputy Director, Health, Environment, Safety and Security Management Systems (HESS-MS) at Holland America Group

A fascinating, page turning reflection of Microsoft's rapid growth history in the '80s and '90s complete with lessons and questions we can all use today no matter the team we lead. This is a book about leadership, culture and values – on the way up, on the plateau and then growing once again. I often wonder how much further along Canada's tech sector would be in 2017 if Dave Jaworski had not followed his opportunities in America.

Scott Baldwin, MBA, ICD.D, Corporate Director

Microsoft® *SECRETS*

AN INSIDER'S VIEW OF
THE ROCKET RIDE FROM
WORST TO FIRST AND LESSONS
LEARNED ON THE JOURNEY

FORMER MICROSOFT EXECUTIVE
DAVE JAWORSKI

NEW YORK

NASHVILLE • MELBOURNE • VANCOUVER

MICROSOFT *SECRETS*

Published in New York, New York, by Morgan James Publishing. Morgan James is a trademark of Morgan James, LLC.
www.MorganJamesPublishing.com

ISBN 978-1-68350-420-7 paperback
ISBN 978-1-68350-421-4 eBook
Library of Congress Control Number: 2017901035

Cover Design by:
John Price
Chris Treccani, 3dogdesign.net

Interior Design by:
Megan Whitney
Creative Ninja Designs
megan@creativeninjadesigns.com

Creating the *FUTURE*

"The key thing is to keep doing more powerful applications, to justify using the personal computer on every desktop or eventually in every home."

— Bill Gates, *Personal Computing*

"The fact that people are getting exposed to computers at such young ages will change the thinking in the field."

— Bill Gates, *Programmers at Work*

"Nobody ever said the computer business would be easy."

— Adam Osborne, *PC World*

"Desktop functionality on a laptop is a challenge, and the sooner Apple has a laptop Mac, the better."

— Bill Gates, *Macworld, 1988*

"I'm a great believer in delegation. The only way you can run a growing company is you must build great managers. I want to have a group of people working for me all of whom would absolutely desire my job."

— Jon Shirley, during his tenure as president of Microsoft

"As computers become more deeply involved in society and reachable by more and more people, they are going to have to become more and more fun."

— Neil Shapiro, *MacUser*

"Think of buying a computer as like buying a car. A car just moves your body; your computer, though, is the chariot of your mind, carrying it through the whole universe. How much is your mind worth to you?"

— Ted Nelson, *Computer Lib*

"If the automobile industry had advanced this fast (optical storage technology) since 1981, we'd have cars that go from 0 to 60 in one second, circled the globe on a cup of gasoline, and cost about half as much as they did six years ago."

— Min S. Yee, Optika 1987 Conference

"There is a reason that the leading edge is often called 'the bleeding edge:' the pioneer users of new technology are often battered and bloody by the time the technology stabilizes."

— Robert R. Wiggins, *MacUser*

CONTENTS

Ten Years That Changed *OUR WORLD*

The time period from 1985 to 1995 will go down as one of the most important times in history—not just technology history but the history of humanity. These ten years mark the period in which technology rose to the point it enjoys today; it is pervasive in every sector of our lives, and not just in North America but all over the globe. That was certainly not the case in 1985.

Consider some of the highlights in technology's history from 1985 to 1995:

- Microsoft went from worst to first. In 1985, we at Microsoft were literally dead last in every software application category. We were number one with MS-DOS, the operating system that powered the IBM PC and compatibles. DOS was the fuel for the entire engine of Microsoft until the other applications really started to contribute. By 1995 we had finally risen to first place in every major category.

- Intel and Microsoft drove the hardware industry forward at a dizzying pace, showing the reality of "Moore's Law" in practice. Gordon Moore of Intel said that the power of the microcomputer chips would double every eighteen months. That they did—and then some.

- Apple staved off being eliminated from the tech industry. Many of its predecessors were not so fortunate. Microsoft helped in two major ways. First, Microsoft created Microsoft Excel, which enabled the Mac to be a solid contributor in businesses. Second, Microsoft infused $150,000,000 into Apple at a time when it desperately needed the funds.

- Microsoft went from being a privately held company with more than $50 million in revenue to a publicly traded company worth billions. Today Microsoft and Apple have valuations larger than do many countries.

- Microsoft successfully refuted a $5.5 billion dollar lawsuit from Apple, opening the way for today's open source software industry.

- Microsoft Office was invented.

- Microsoft's vision of "A microcomputer on every desk and in every home running Microsoft software" went from a crazy dream to within reach.

So much more happened that still impacts us today and will impact generations to come.

You might wonder why the above list focuses so much on Microsoft. Weren't there other players, other innovators? Yes, of course there were. But during that ten-year period, Microsoft was the company that drove the hardest, built the fastest. And I was there during this rapid rise to the top. I kept meticulous notes and took lots of photos. I documented the risks we took, the dreams we shared, the lessons learned, the hopes realized, and the mistakes we made. Many of the issues we faced at the time are similar to issues confronting leaders in business today. All of us can learn from Microsoft's past. Were the approaches Microsoft took the right ones? What would you have done? What is happening today that warrants a similar approach? We can either learn from history or repeat its mistakes. But before we can learn from Microsoft's past, we need to get the history right. That's one of my goals for the following pages.

A great deal has been written about Steve Jobs and Apple. A good amount was written during Steve's lifetime, and movies and documentaries have been produced on his life and leadership of Apple. Not nearly as much has been produced about Bill Gates and Microsoft, especially in the time period that I was at Microsoft. Walter Isaacson's book *The Innovators: How a Group of Hackers, Geniuses, and Geeks Created the Digital Revolution* spent most of its time looking at Bill in the period before my time at Microsoft, and Isaacson characterized him in a way quite different from the person I came to know. The story of Microsoft needs an insider's portrayal and that's what I give here.

I am as accurate and careful in my telling as I can be. You will find exhilarating stories and exasperating ones. Some of the stories are not flattering, at times not to me and sometimes not to Microsoft. Yet I believe they need to be shared. For from them we can glean insights and the principles on which they are based, and these, I believe, absolutely matter to the decisions being made today and in the years to come. We need guideposts for the journey, both personally and professionally. Through Microsoft's rocket-rise in the technology field, we can locate many of these guideposts that can light our way too.

Among the stories I share, several of them are familiar to family, friends, and others I have worked with over the years. Yet, at the same time, many of the "facts" you will find elsewhere are simply incorrect. This book is an attempt to correct some of what now stands as the historical record.

I have not been itching to write this book. In fact, for many years I was actually against that idea. I thought that such an effort would likely appear vain. And if the book was just about me or for me, I definitely had no interest in creating it. My passion is helping people and organizations grow and achieve their full potential. If any book I created did not move that passion forward, I would not produce it.

What I found, however, is that as I shared some of my Microsoft stories, insights arose and listeners showed great interest in them. Many fascinating discussions about leadership and principles in business started with the telling of one of those stories. The issues and discussions were clearly relevant to business and personal life today. Because of experiences like these, I have been urged for quite some time to put in writing my time at Microsoft. In one thirty-day period I was asked questions about writing a book so many times that I decided it was time that I share my account of what happened and why.

As you read what follows, you will find that not every decision we made at Microsoft was either clearly right or clearly wrong. I trust you will find yourself asking, even as you reflect on your own situation, where are the lines for what you and your company should and should not do? When should they be crossed, or should they ever?

I will also be telling you some secrets—some only my family knew. Some of these secrets were known to only a handful of people within the company at a time when it grew to over fourteen thousand employees. Like the secret recipe for Coca-Cola or Colonel Sanders' chicken recipe, these secrets were literally changing the competitive landscape in the technology industry and were rewriting the business rules of the day. Understanding these secrets and the thinking behind them can provide strategic insights and advantages to you and your business. Better still, they can help you define your own secrets to accelerate you past competitors and over hurdles to success.

One of my hopes for this book is that a new generation of Microsoft employees will benefit from this firsthand account of their heritage, learn from it, and make Microsoft great again. Those of you who are fans of Apple, Google, Amazon, or (put your favorite technology company here) may not like that last sentence. But I believe that competition is great for everyone—for consumers, the industry, even the world. It drives creativity and innovation to new heights. "A rising tide floats all boats." I have found competition to be like that. There was a period where Microsoft dominated and all other tech companies seemed anemic. That was not good for Microsoft or anyone else. Competition pushes us to solve real problems sooner. It searches for differentiation and, in the process, results in breakthroughs.

Noticeably missing from my story is Microsoft's co-founder, Paul Allen. I have only met Paul a few times and no more than a casual hello was exchanged. Paul became very ill and left day-to-day life at Microsoft in 1982, a few years before I joined the company. He remained on the Board of Directors through 2000. For my record of Microsoft's history, then, I never knew Paul well enough to include him.

In some ways, to me at least, the story of Microsoft seems epic, like Tolkein's *Lord of the Rings*. It is a story about …

Leadership

Risk

Adventure

Controversy

Battle (with Apple, Google, the US government, and others)

Fall from grace

Return of the king (and the crowning of a new king)

I'd love to see director Peter Jackson's version of the Microsoft epic. (And so would Susan, my movie-loving bride.)

As you read, you will need to judge for yourself what I have said here. I also urge you to take notes. Engage with other readers and with me on social media (@DaveJaworski on Twitter). Discuss and debate the points I make in this book. My hope is that *Microsoft Secrets* will play a role in helping us all build better companies and better lives that propel us forward on each of our next rocket rides.

One more thing. Throughout the book, you will find a section called "Take Action!" I believe that the lessons I learned along my journey can help you have a better personal life, professional life, or both. So I hope you will consider this section as a place where you can turn my lessons into actions that will benefit you.

Join me now on the rocket ride journey that has changed all of our lives. I present it to you as a journey through history, a lesson for today, and my personal journey that melded my faith, a new-found love of technology, and passion for helping people achieve their goals.

Part One
A PERSONAL JOURNEY

ONE

Oceanography to *COMPUTER SCIENCE*

My first love was God. My second was music. My third was water.

Jacques Cousteau was a famous underwater explorer and innovator. His stories inspired me as a boy. The idea of engaging with the animals of the oceans, especially whales, enthralled me. So I decided I would pursue a career in oceanography.

Swimming lessons had taught me how to help someone who is drowning, how to breathe properly for longer periods under water, and many other "life skills" that I first thought only applied to my time in the water. Lessons can be learned at all points of our life journey, and we can glean many from the journeys of those who have gone before us. Jacques' adventures as well as my own lessons "in the water" taught me how to deal with unexpected and even life-threatening situations; how to handle my emotions and make decisions in times of uncertainty and rapid change; how to pioneer in unchartered waters; and how to persevere in the face of failed experiments. These lessons would prove useful in ways I never imagined at the time.

St. Paul's High School, a Jesuit school located in Winnipeg, Manitoba Canada, provided a formal education and a principled foundation for my personal and professional life. The education I received and friends made during those years impact me to this day. Two technology breakthroughs were introduced to me at St. Paul's and made a significant impression on me. One was the electronic calculator. This handheld device could be programmed and could do so much more than simple math. The second was the computer. A small group of students under the leadership of the school's math wizard professor, Fr. Leslie Marosfalvy, were able to use this new technology. Special access was required to even enter the computer room. It was guarded and remained behind locked doors, which added mystery and intrigue to the machines the room housed. I was fascinated by the fact that these amazing machines could run processes that a person's mind conceived. I did not take the computer programming class because I incorrectly assumed only math wizards could apply. Yet the seed planted would germinate almost immediately after I graduated from St. Paul's.

Nevertheless, the call of the sea still beckoned this prairie boy. In order to get a degree in oceanography, I first needed to take a general science degree. So that was the next step on my journey. On my first day at the University of Manitoba, freshmen had the opportunity to take optional tours. Since my curiosity had been peaked by the work being done at St. Paul's with computers, I checked the box to take the optional computer science tour. While the machines were impressive, it was the work of the masters students that blew my mind. All their projects focused on helping people with permanent or temporary disabilities. This amazing new technology was going to improve the quality of people's lives, in some cases dramatically. That enthralled and moved me. It still does.

Still believing I was destined to be the next Jacques Cousteau, I started down the path of the science degree. I also completed my swimming training and joined a program to get certified to use scuba gear for diving. Jacques Cousteau was waiting for me in the oceans. Or so I thought.

Just a few steps out of the starting block, this career plan took an unexpected turn. I had a minor head cold during one of my early scuba dives. I couldn't clear my sinuses to relieve the pressure that naturally builds as you go deeper in the water. Pressure kept building in my head until it felt like my head would explode. The headache that resulted from that dive was unlike any I'd ever experienced before. This became a recurring theme on my dives, and I learned that they were due to narrow sinus cavities. This physical condition and what it caused led me to part company with my dream to be like Jacques. Signing up for a life of extreme pain was not something I was willing to do. Nor did I crave an oceanography desk job. It had to be deep-sea diving or nothing.

A trip to the university's administration office was in order. It was time to change the declared major for my degree and that became computer science. Learning how to use technology to improve people's lives was now my life mission.

At that time, though, computer science was all about punch cards. That's right. No computer terminals. Just a deck of cards that had holes punched in them. Each card represented one line of code. Those cards were read into a card reader, which then fed the instructions to the university's Amdahl Mainframe computer. Then you would walk around to the printer where the output from your job would appear. Did the program work? Did it output what you asked it to do? Did a card get bent in the card reader? If the program failed or a card (or two) were damaged, you had to get back in line to wait for a card reader to become available again so you could restart the process.

My graduation certificate declared that I received a bachelor of science, majoring in computer science. Just one year later any such graduate's diploma read "Bachelor of Computer Science." Additionally, every class that followed us was able to use terminals to interact directly with the mainframe. No more cards. And soon after, students would receive online and almost immediate feedback about their programs. Nice. I missed all this by just one year. Still, my undergraduate study in computer science at the University of Manitoba had two standouts: the Logic 100-level course and great teachers.

Computers are all about logic and breaking problems down into small steps. Logic 101 taught me that the logic flow was the main skill and that programming languages was about semantics. In other words, whether your thinking was correct—the logic—was relevant across all programming languages. The language was not important. Some languages were more efficient for certain types of problem-solving. Some offered data structures to help move information more easily for the programmer. Whether we coded in Fortran, BASIC, APL, COBOL, PL1, or any other language, it was the quality of your thinking—the logic that you applied—that mattered most. The specific language was about semantics and "grammar" or structure. The underlying logic determined if, once you had all the correct syntax, your program would produce the desired result. You needed to communicate with the right "vocabulary" and the right "grammar" to be understood. The logic would determine whether or not you could get to your goal.

Some people get religious about programming languages. While every computer language has various pros and cons, I believe that the underlying logic you use is what differentiates great programming from average or poor programming. In this sense, the language used is not what matters most. Computers are precise and require syntax to be correct. And there are efficient and inefficient ways to write code. Yet, at the end of the day, it is your underlying logic that is critical to failure or success. This is the case in business as well. Logic and flow or process matter more than semantics. People get religious and political around semantics when often the issues and opportunities lie in the logical approaches taken by the business.

Great teachers matter too. A lot! I was blessed with many wonderful teachers in grade school and high school. Yet none rocked my world like my university computer science professor, Lorne McMillan. Lorne had a way of teaching and communicating that made hard problems easy and brought clarity to previously muddy areas. And he did so in an incredibly personable way. In anything I was to teach in the future, I wanted to do so like Lorne had done for me. I value the skills of people who can teach. I also learned the incredible importance of a continued investment in training and development at Microsoft. More on that to come.

My revised priorities were now God first, music second, and technology third. For the next year I moved music into the top career path goal and placed computer science into the backup position. Learning how to leverage technology to improve people's lives now drove my academic studies. However, while completing my degree, I worked nights and weekends to develop a career path in music that I hoped would lead to writing and recording songs.

I grew up in a very musical home. Mom taught accordion and played beautifully, and Dad could play piano or guitar by ear. They taught me about music. Not a genre. All music. They would go from playing country's Charlie Pride to rock's Alice Cooper. They took me and my brothers to see in concert such notables as Elvis Presley, Sammy Davis Jr., and Mac Davis. I gained an incredible respect for great music, no matter what the genre.

Guitar playing and writing music grabbed a piece of my soul. I played acoustic guitar with the folk group at church, starting in the group at the tender age of eleven. I went on to play more than thirty years in church. The folk group became a rock band and then a praise-and-worship group. All I knew was it was music for God with the goal of getting people in church to relax and sing out freely to Him. It

The Band: from left to right, Gordon Marce, Wayne Jaworski, John Deegan, Phil Paolucci, and me

was here that my number one passion (God) and number two passion (music) were joined at the hip. Playing music in church was not to be a performance. It was instead a spiritual experience, one I clearly felt. I also felt it when I played with my brothers and friends in bands outside of church settings.

The music program at the university seemed to be all about theory and, in my mind, unnecessary for someone who wanted to write and produce music. Yet my parents had stressed the value of getting a university education. So, even though computer science was my major, the career plan, post Jacques Cousteau, was music—writing and producing music—and my fall back plan was the computer science degree. Once I made the transition to studying computer science, I managed to keep my love for music in my life. I secured one of the four paid positions at the university's radio station, CJUM-FM. I also moved on to a commercial station, actually two. I went to work for Armadale Communications. On the AM side, I was a country music DJ at CKRC in Winnipeg. I also worked the company's adult easy listening station, CKWG FM. I lived up to my initials. I was DJ the DJ.

At CKRC and CKWG, I met wonderful people and broken people. One of the broken ones, a fellow DJ, shared how he was in his fifteenth job and seventh city in his career. He talked about how the station general manager (GM) would change and then the new GM would bring in "their people," which meant that the current DJs would lose their jobs. So this man I knew would pick up his life and move to the next best opportunity, often in another city. He managed to become successful in the business world, having secured one of the better time slots for his on-air work at the station. But his success and moves had cost him some failed marriages. He was quite unhappy.

One Saturday afternoon I was asked to do a "remote," which meant going out to a location and having the on-air announcer connect with me during their show while I showcased the advertiser or event. In this case, the event was a fundraiser for the Lioness Club in Winnipeg. They were doing a twenty-four-hour bowl-a-thon. A young lady named Susan was tasked with educating me on the event so I could talk about it intelligently on air. When she finished her talking points, I realized I had heard almost nothing and was totally mesmerized. I asked her if she had a boyfriend. Her reply, "Yes, I'm sorry, I do." I heard the "I'm sorry" as if it was light beaming in through a door. She was breaking up with her boyfriend, and she wanted to complete that process before starting another relationship. I was eager to be that next relationship.

That day marked the beginning of our life together. After several months of dating and an engagement period, we got married. We have been together ever since, married for thirty-four years now. And all of this started just thirty-six years ago, on the day of doing a remote for radio at a local fundraiser.

With the unhappy lives and broken dreams and marriages of my radio friends, my new relationship with Susan, the desire to provide a great life for us together, and the growing passion I felt for technology and what it was going to do to improve people's lives, I made the decision to switch my vocation and my avocation. Computer science became my career focus. Music became my avocation.

Susan and I were married in July 1982. On our honeymoon we purchased an Apple II Plus computer. Yes, we were that geeky. We had as much fun playing with the computer and learning it as we did touring Disneyland on our honeymoon. Disney and computing are still our constant companions.

TWO

Bit By The COMPUTER BUG

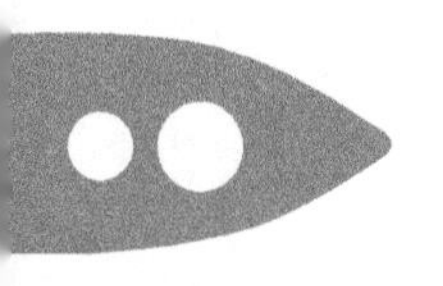

The computer science degree first went to work at Great-West Life Assurance Company, programming their mainframe computers. The degree wasn't even complete when I started at GWL.

Sitting in a cubicle and interacting with a computer screen all day got old quickly. Interacting with others, however, provided me with energy. Still, the cubicle drained even that. So I switched my focus from writing code to teaching people how to write for the mainframe.

Microcomputers, specifically the Apple II series, the IBM PC, and the Apple Lisa—Apple's first computer with a graphical user interface—provided immediate feedback and were considered "personal" computer devices. These "micros" started making their way into the company. The reality of these devices making our lives better "bit me" (Apple pun intended). I saw great potential ahead.

In time, an Apple II with Visidex replaced all the paper on my desk. My notes went into Visidex. When someone would walk up and ask a question, I could call up almost any note with a few keystrokes.

As more and more microcomputers started coming into the company, someone was needed to teach people how to use them. Given the uses I had already demonstrated, I was nominated.

Citation Software

After three years at Great-West Life, I decided to focus all my energy on the emerging field of microcomputers. I joined Citation Software, which at that time was Canada's largest distributor of microcomputer software and hardware peripherals. They handled over eighteen hundred products and ninety product lines. I joined the technical support team, providing support for dealers and end users across Canada. I was promoted to lead tech support when my manager left the company.

We had a lot of companies and products to support. I quickly saw that the most common problems people experienced were the same. So I grabbed an unused computer—a DEC Rainbow—and installed the dBase II database program. We logged the problems people reported as well as the solutions, step-by-step. When anyone called in with an issue, we would first search keywords in the database to see if the problem had been previously logged. If it had, and we had a documented solution, we'd ask if the client had a fax machine. In the pre-email days, almost everyone had or had access to a fax machine. We would send our client the instructions, telling them to call back if they needed more assistance. Then we could move on to the next caller. Two of us were able to handle as many support issues as teams over five times our size. Leveraging technology to accelerate and automate business processes was an important lesson that I have repeated many times in my career.

Systems create leverage. If you are not creating systems in your business, you are missing opportunities to use your resources in more effective ways, to increase your competitive advantage, and to add speed to your ability to respond to changing conditions.

Citation Software's top executives started to make decisions that hurt the business. They did this based on the assumption that the customer was always right. To them this meant purchasing products that only one customer wanted and also giving a 100-percent return policy on it. Customers often took advantage of this policy, even for expensive products like plotters. These decisions, done with good service intentions, killed the business. Another lesson learned: what is right for the customer is not always right for the health of the business.

The next move in my career path turned out to be an especially important one. I knew microcomputers were going to change our lives. I needed to pick who I thought was going to lead this revolution.

- Are you leveraging technology to create strategic advantages for your business? For your personal life? If not, why not?

- Are you doing things for your customer that are putting your business at risk? Can you eliminate these actions or change them to create a win-win?

THREE

Apple Or MICROSOFT

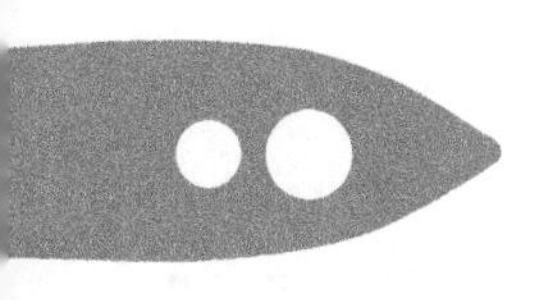

"I saw your résumé in *The Globe and Mail!*"

What?! That was like telling someone in the USA that you saw their résumé in the *Wall Street Journal. The Globe and Mail* was Canada's WSJ. Henry Quan, a friend who had worked with me at Citation and now led marketing at another tech company, was the voice on the other end of the phone. "Henry, what are you talking about?" I asked.

Two other people called me that day. One said the same thing as Henry, almost verbatim. The other explained that he'd seen a job posting that read like my résumé. Henry said the same thing in his answer: "Dave, this is perfect for you."

KPMG conducted the job search for the company. They described the position to me in great detail. They said it would be at my expense to fly to Toronto from Winnipeg for the interview. I then asked them to review my résumé and let me know if I would need to pay for the trip or if they would cover my expenses. Microsoft decided to bear the cost. What tipped the scale was what they kept hearing from other candidates. Candidates were asked to describe great technical support. Several said, "Dave Jaworski." Little did they realize they were setting the stage for me as a candidate for the position. The recruiters, hearing that I was an applicant, decided it was worthwhile to bring me out to talk at their expense.

I flew out to Toronto on January 30, 1985, and my job interview with Microsoft began at 10 a.m. the next day. From that moment on, the direction of my life and that of my family changed considerably. Henry had been right. The job fit my résumé perfectly—except that it was not the job I wanted.

Job Offers: Microsoft and Apple

I had been hooked by Apple from the time Susan and I bought the Apple II Plus on our honeymoon. My experience using the same types of Apple II computers and then Apple's Lisa at Great-West Life turned me into an even greater fan. Apple simply made wonderful personal computers. By the time Henry had called me, I had already reached out and applied to Apple for a training job in their Orange County, California office. Apple had extended an offer to me at the same time Microsoft called me to meet in Toronto.

Bob O'Rear and Rich Macintosh interviewed me for the Microsoft job. Bob is one of the original eleven people pictured in the historic

Microsoft first-company picture. Bob is one of the nicest people you will ever meet, and he is one of the smartest. He is also the humblest person I know. He is a true rocket scientist. Prior to joining Microsoft, Bob worked at NASA, and his work was the difference that kept the early astronauts safe and alive.

Microsoft's Original 11: Bob O'Rear is just above Bill Gates (bottom left)

Rich MacIntosh was new to Microsoft. As general manager, Rich would lead the new Microsoft office opening in Canada for the first time. He had been an executive with Computer Innovations, a Canadian computer retailer. Rich is brilliant. To this day, I credit him with being the greatest unsung hero in the Microsoft story. He stayed behind the scenes to most people in the industry and the millions of people who benefit as users of technology. Yet his leadership influenced Bill Gates, Jon Shirley, and many other executives at Microsoft. He built a team that not only handled the hyper-growth period at Microsoft but helped generate that growth, moving the company from dead last to first place among tech companies. It all happened during his tenure. Rich taught us many invaluable lessons, many of which I have captured in this book.

Now, back to the job interview. Rich and Bob concluded my job interview with two offers for me to consider: Either join Microsoft Canada as head of tech support, or join Microsoft Canada as a marketing representative.

"I'll take the marketing rep position," I responded.

Rich replied, "Think about it and let us know." He reminded me that the tech support manager position came with an office, a company car, and a team of people reporting to me. The marketing rep position required extensive travel, provided a car allowance (I needed to provide the car), and a cubicle versus an office.

"I'll take the marketing rep position," I repeated.

Rich asked why I was so adamant about this position over the tech support one. I shared that I wanted to put my tech knowledge to work at Microsoft by helping people learn how these new computers and software could improve their lives. I didn't want to continue doing the kind of job I had at Citation Software.

Rich replied, "Think about it and let us know tomorrow."

First thing the next morning I made two calls. The first was to Rich at Microsoft. "I'll take the marketing representative position," I told him. My next call was to Apple in California, letting the company know I would not be taking the job with them.

Why did I choose Microsoft over Apple? Because I believed that software drives hardware. I still believe that. No matter how beautiful hardware is designed, it is the software that gives it its brains and enables it to function. Beautiful hardware without beautiful software feels clunky. At this time in Microsoft's history, Bill Gates had already espoused another key tenet for his computer systems: they would have a common user interface across many products that would make it easier for people to learn and use software. This approach started with common menu structures and later extended into common mouse and interface interactions.

Apple's Steve Jobs believed hardware drove everything and that software was not as important. He said as much in multiple public presentations. He later "changed history" with one of his many "reality distortion field" moments, saying he always believed in the combination of hardware and software as the key to great computing experiences.

Given Steve's original stated directions and Bill's, along with my beliefs in the value of software, I chose to join Microsoft.

Employee #3 at Microsoft Canada

My first official day as a member of Microsoft Canada was March 1, 1985. I was employee number 3. Though just twenty-four years old at the time, I didn't feel too young to be in such a role. After all, Bill Gates was only twenty-nine. I was Microsoft's 735th employee in the world. The company was still privately held and had achieved over $90 million in worldwide revenue. (Microsoft did not go public until 1986.)

Microsoft Canada started its operations in a small one-room office above a pancake house on Airport Road in Mississauga, Ontario. We met and planned for the opening in that office with regular visits downstairs for "fuel." Then, in May 1985, we officially opened the doors for business at Microsoft Canada at 6300 Northwest Drive, near the Toronto airport. Our critical mass at that point was just seven of us. By June 1985 we had grown to a full-time staff of fifteen, and by the beginning of 1988, we numbered twenty-eight in Toronto. We also expanded to have offices in Vancouver, Calgary, Ottawa, and Montreal. We were on the move. By the time I first left Microsoft in 1993, we had reached over fourteen thousand employees in the entire company. (I returned a year later for an additional two years on the product development side of the house.)

Microsoft Canada

By 1990 we were generating over $1 billion in revenues. It was up to $6 billion by 1995. Microsoft was booming.

In my first weeks as a marketing representative, I was assigned half of Canada from Toronto to Vancouver. A little while later I was asked to cover all of Canada since my counterpart quit. I hit the road almost immediately, living coast to coast and covering Canada's five-and-a-half time zones.

Soon after my start at Microsoft Canada, the US sales team invited me to attend their sales meeting. There was a total of thirty of us. One of our goals was to get personal computers running Microsoft software into corporate America. It was thirty of us versus thirty thousand IBM sales people. If IBM chose to put a few people on us to smother us with attention, they could keep us from our goal. It was a true David-versus-Goliath situation. And David ended up winning this high-tech sales battle.

Windows 1.0

Microsoft Canada achieved many firsts, some which were simply the benefit of scheduling. For example, the Canadian Computer Show was scheduled for November 18–21, 1985, just a few days ahead of COMDEX, the major US trade show for the technology industry. COMDEX was entering its fifth year and was held November 20–24.

Just months after we opened our doors in November 1985, I had my first meeting with Bill Gates. Bill came to Microsoft Canada and the Canadian Computer Show for the first release of Microsoft Windows 1.0 on November 18, 1985. Two days later, Windows 1.0 was released and shown at COMDEX in the USA. Microsoft Excel 1.0

for the Mac had been released on September 30, 1985 and was featured in the booth banners at the show (see the picture below). (A little trivia: there never was an Excel 1.0 on the PC. When Microsoft released Excel for Windows, it started at version 2.2, synchronizing version numbers with the Mac version that was out at that time.)

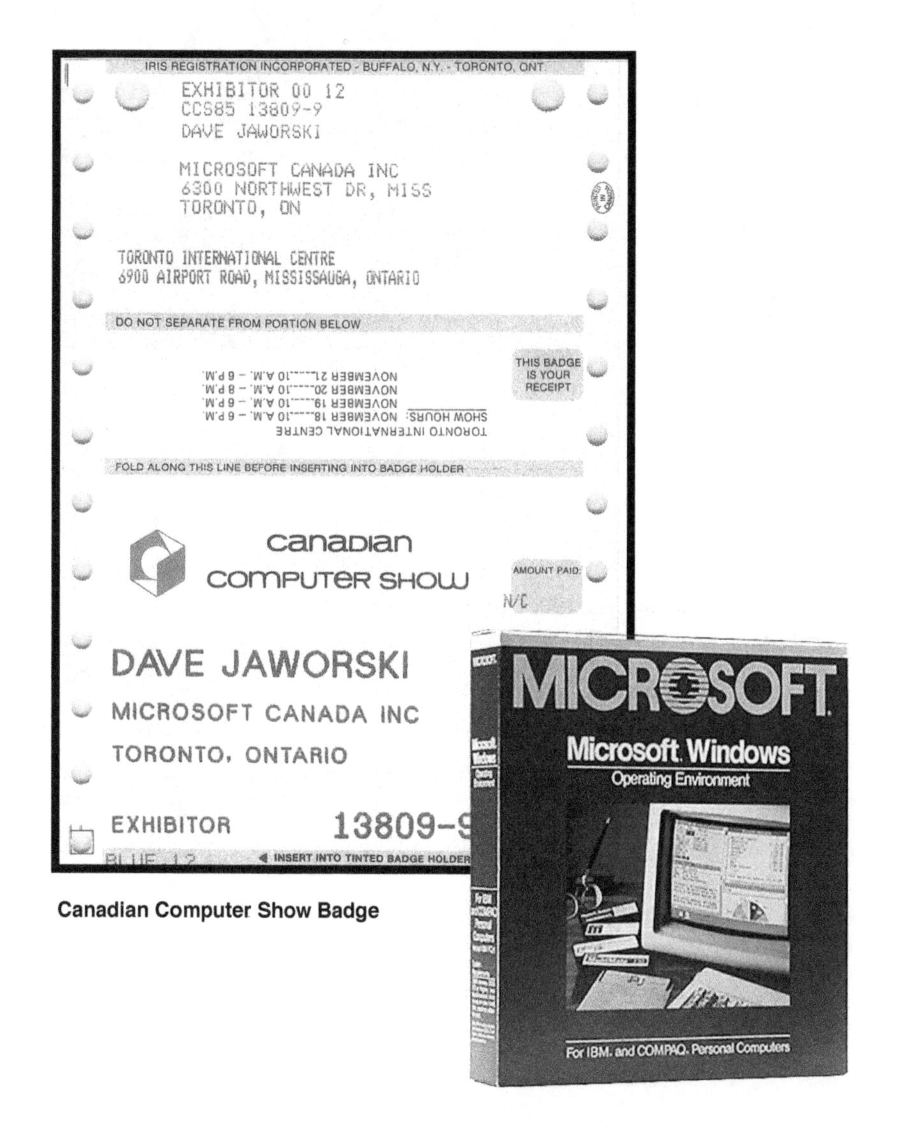

Canadian Computer Show Badge

From the right of the booth, Louise Martel, Bill Gates, me, and Rich Macintosh at the Canadian Computer Show, November 18, 1985.

Everything I had heard about Bill Gates proved to be true in person. He was smart and visionary. He had an inner drive that is still felt today. He had high standards and expected greatness from his team. He was a quick study. He asked a lot of questions to get to the heart of your thinking as soon as possible. He had strong opinions yet remained open to other possibilities, and he stayed curious in all matters. All of these traits served him and our company well.

So began the rocket ride. The years that followed were filled with lessons learned.

Part Two

PRINCIPLES MATTER AND OTHER LESSONS FROM THE ROCKET RIDE

FOUR

The Power of a *SHARED VISION*

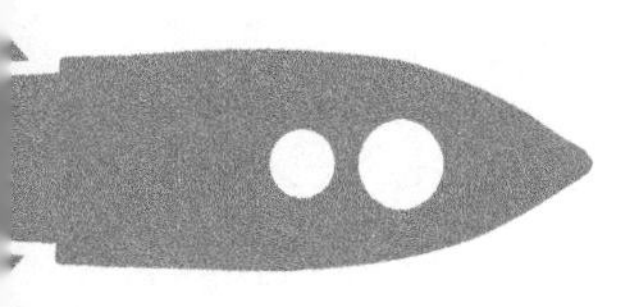

Bill Gates had just turned thirty when I first met him. He was super smart and had great vision. To this Canadian boy, he was the Wayne Gretzky of technology. Wayne's famous saying was: "A good hockey player plays where the puck is. A great hockey player plays where the puck is going to be." Bill Gates led the company to where the technology puck was going. He did that for Microsoft and the technology world in general.

Bill articulated the Microsoft Vision as "A microcomputer on every desk and in every home running Microsoft software." The Microsoft Vision drove the entire company forward. Virtually everyone you met in the company could articulate it. It was a shared vision. Author Jim Collins talks about having your "BHAG"— Big Hairy

Audacious Goal. "A microcomputer on every desk and in every home running Microsoft software" was our BHAG. At that time few people in the general public had used a computer. They had heard of them, yet many had not even seen one in person.

When you have a vision, not everyone will get it. That's okay. Go forward with it anyway.

Today, of course, the idea of a microcomputer on every desk and in every home is easy to visualize. Even young children know how to operate iPhones and iPads, swiping and touching to navigate. Our three-year-old grandson Elijah looks puzzled when an adult cannot find and start Spotify to play music. He motions for the iPad or iPhone and moves quickly through the icons and folders to locate the familiar icon, start the program, and navigate to the music he wants to play. But when I began with Microsoft, today's reality was just a dream.

Apple, Tandy, Commodore, Kaypro, and others had computers for hobbyists. IBM joined in with its personal computer in 1981, introducing the IBM PC. The graphical user interface was first released by Xerox with the "Star" and then made more affordable by Apple with the release of the "Lisa" in January 1983 and then the Macintosh in 1984. These were the pioneering days of personal computing.

Within a few years, what then was considered "portable" computers were available. When flying across the country, almost every person who passed me on the way to use the bathroom would stop and ask, "What is that?" pointing at my portable, luggable computer. (These computers were very heavy and awkward in their earliest implementations.) Today, your phone has more power than those early computers had, and almost every person on the plane has one or more computing devices with them, including smartphones, tablets, and smaller, lightweight computers.

Even devices like Mice were not commonly understood. In fact, as Microsoft Canada started to bring product into Canada, our Microsoft Mouse shipment was quarantined by the Canadian Department of Agriculture. After the shipment's four weeks in solitary confinement, Rich MacIntosh received a call stating that he could retrieve his mice.

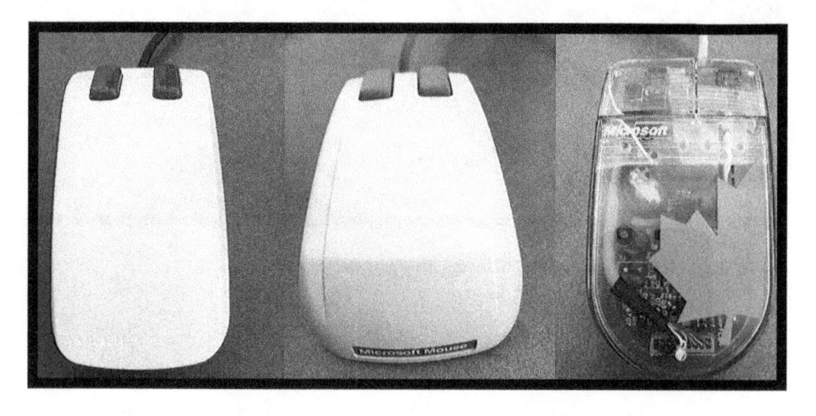

Evolution of a species: early versions of the Microsoft Mouse, including the special Canadian edition

The power of a shared vision cannot be emphasized enough. I believe this truly differentiated Microsoft from most of our competitors. We knew where we were going. We all knew the vision. And Bill ensured the vision, and the ways in which we would advance it were kept in front of us all. We would even be tested on the vision and other product strategies at our national sales meetings. It was not an accident that we all knew the Microsoft Vision.

Training and testing at our national sales meeting.

The power of having a shared vision led me to develop my own personal mission statement. On a birthday many years ago, I took the time to write it down. Over the years I have added sons-in-law and grandchildren to it. Other than that, almost nothing has changed. Here is my personal mission statement:

> I believe in Jesus Christ as my personal Lord and Savior. My life is a journey to eternal life with God. I value my relationship with God more than anything else in my life. I value everything that supports my journey to eternal life with God. I want to follow God's will and plan for my life. I am a servant of God.
>
> I care deeply about my marriage union with Susan. She is God's gift to me. I want Susan to be at my side to present me

to God, and I to present her to God, when our time on earth is done. I enjoy spending time with Susan, ever growing our relationship together and with God.

I value the gift God has given Susan and me in our children and our grandchildren! I value our relationship and want it to always be one of love and open communication. I want our children and their children to know Jesus and to be with Him at the end of their time on earth. I want our children and their children to have a lifetime of opportunities through their positive approach to each day.

My work reflects my values. In work I value the committed pursuit of excellence and strive for excellence in all I do. I value, in all aspects of life, proper moral and ethical conduct. I enjoy work that gives me the opportunity to make a difference with my life, versus just filling a position like a cog in life's wheel. I enjoy my work. Work is and should be fun. I enjoy using the talents God has given me in all that I do. I value creativity and pray for guidance so that all I create is a reflection of God's will. I want my work to lead and inspire others to their greatest potential and to God.

My personal mission statement has been a powerful guide for me on life's journey. It brings me back to a focal point. It clarifies my "why" when life gets busy. It helps me prioritize decisions and action steps. A business mission statement does the same thing. A well-defined mission statement will help you make decisions. It will ensure everyone is on the same page. When business life gets busy, it can bring much needed clarity.

- Do you have a clear and concise vision for your team?
- Does everyone on the team know it? (I mean every single person at any level in your organization.)
- Do they understand it? How do you know?
- Do you have a personal mission statement?
- Take the time to give yourself this wonderful gift.

Email Address on a Business Card

From the beginning of my days at Microsoft Canada, I included an email address on my business card. (The Envoy ID was the email address.)

Microsoft Canada Inc.
6300 Northwest Drive
Mississauga, Ontario
L4V 1J7

Tel (416) 673-7638
Telex 06-968547
Fax (416) 673-9728
Envoy: D.J. Jaworski

Dave Jaworski, B.Sc. (Comp. Sc.)
National Sales Manager

I soon included multiple email addresses, because the systems in operation could not yet cross-communicate. In this later business card, Compuserve, AppleLink, and Prodigy addresses all enabled people to send electronic mail to me.

Microsoft Corporation
One Microsoft Way
Redmond, WA 98052-6399

Tel 206 882 8080
Fax 206 883 8101
Compuserve 73075,1243
AppleLink D2394
Prodigy HVVN24A

Microsoft

David Jaworski
General Manager
Sales Operations
USSMD

As you can see, business cards then included Telex addresses for communication. Today, most people have not even heard of Telex machines. Telex systems used a different communication system for sending messages. Much like a fax message is different than an email message, Telex machines were dedicated and yet another unique communication tool.

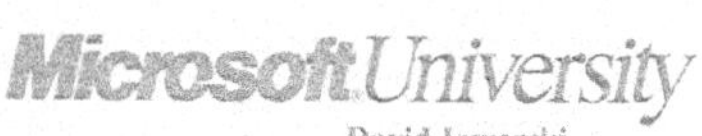

David Jaworski
General Manager, Microsoft University
SMSD
Tel 206 828 1515

Mailing Address MSU
10700 Northup Way
Bellevue, Washington
98004-1447

Corporate Office
206 882 8080

Registration
206 828 1507

Telex 160520
Microsoft BVUE
Fax 206 822 1429

It was a radical idea to use computers as communication tools. That was a big part of the vision that captured me for the use of technology. And to me, it made great sense. Yet even in 1993, having your email address on your Microsoft business card was still not standard for all employees as evidenced by the "standard issue" Microsoft University card I received at MSU.

Today, your email address is probably more important than your physical mailing address. In fact, many business cards include just the person's name, phone, and email address. Some add social media like Facebook, Twitter, and LinkedIn, even before they add a physical address. This shows how that early radical idea has become the new norm. Communicating with computers? Absolutely! It happens every day, in every country, at virtually every level of society, and in every sphere imaginable—to name just a few, academia, business, personal, family, military, politics, government, utility companies, and even the recreation industry. Computers have changed our lives and have become indispensable communication tools.

- Review how you let your customers communicate with you and your organization. Are you letting them communicate in the ways that are most effective for today's emerging technologies? Would your business benefit from taking a leadership role in new communication channels?

FIVE

Steve Jobs and a BUSINESS CARD

A lot of books have been written about Steve Jobs. He has become a folk hero. He was certainly a brilliant person. His contributions are admired because he has positively impacted more industries than almost anyone else in our day. Few people have had the impact Steve has on the technology industry. And the technology industry has effectively impacted every industry sector. Additionally, he has impacted the music industry and the movie/animation industry.

As great as the accomplishments have been, Steve's method for creating that innovation included him running roughshod over many people within his own companies. When the book *iCon Steve Jobs: The Greatest Second Act in the History of Business* was released in 2006, it painted a truthful yet unflattering view of Steve Jobs. He immediately had the publisher's entire line of books removed from all Apple Retail stores. Other books, including the Walter Isaacson biography, painted similar stories of Steve. Of late, a new crop of books and articles have tried to recast Steve as a gentler and kinder person. Even employees, including reknowned Apple executive and designer Jonathan Ive, have said, "If he was that much of a jerk, I would not have worked with him." Books such as *Becoming Steve Jobs: The Evolution of a Reckless Upstart into a Visionary Leader* look to dispel the Walter Isaacson biography.

I have no axe to grind with Steve. I too admire his contributions. On the other hand, my interactions with him along with the firsthand interactions of some of my close friends who worked at Apple support several character issues that iCon and the Isaacson story shared.

For example, I was at the Los Angeles Bonaventure Hotel for a BusinessLand event in 1990. I had never met Steve Jobs in person until that day. Steve came across the lobby and saw the Microsoft logo on my shirt. He redirected himself to come charging at me and started yelling to ensure an audience would pay attention to the charade. "Hey!" he barked at me. "I want to ask you a few questions!"

I replied with great joy. "Steve, it's great to meet you! My wife and I have been huge fans of Apple for many years. In fact, we bought an Apple II Plus on our honeymoon!"

Without missing a beat, Steve redirected himself into the main meeting hall. I was not going to be a successful ruse for his PR stunt.

Around that same time period, my friend Walt Wilson was working at Apple. Walt brought in the CEO of AMR Corporation, parent company of American Airlines, to meet with Steve Jobs in Cupertino at Apple headquarters. AMR was Apple's largest corporate account in the world. As the meeting started, the CEO extended his business card to Steve and placed it in front of him on the table they had gathered around. Steve, without picking up the card, looked at it, put his first finger to his thumb next to the card, and flicked the card across the room onto the floor. Walt was stunned.

The CEO was also surprised. "Why did you do that?" he asked.

Steve replied, "Any card without an email address on it is of no use to me."

Ouch.

In 2005 at the *Wall Street Journal's* "All Things Digital—D5 Conference," I talked with Phil Schiller, head of Marketing for Apple. I had first met Phil when he was an executive at Macromedia and my team was developing software using Macromedia Director and the Lingo programming language it offered. "What is it like working with Steve now?" I asked. "Is he still as intense as he was earlier in his career?"

Phil offered that Steve was intense, although the years had softened that intensity, making him easier to work with. Perhaps that is why Jonathan Ive felt the criticisms of Steve were overly harsh. He simply encountered Steve at a later time in his life when some of the rough edges had been softened.

One of the finest recollections of the intense and, sadly, even mean-spirited Steve Jobs was shared at the DENT 2015 Conference when Daniel Kottke was interviewed by Ellen Petry Leanse. Daniel is an Apple veteran and one of Steve's first close friends. Ellen is also a veteran of early Apple who went on to have significant success in Silicon Valley working with several companies, including Google, Facebook, and Oracle. Daniel has spoken on Apple before, yet this was the first time he had publicly come out to talk in depth about his relationship with Steve. Dan had met Steve at Reed College, a private liberal arts college in Portland, Oregon. Both individuals were eighteen years old, and they tripped on acid together, traveled to India together, and thought through life's meaning together. Dan went on to build parts of the logic board and all the prototypes for the original Macintosh computers. He was a key part of that original Macintosh team. His signature lives on the inside cases of those first Macs.

Dan talked about the books that influenced him and Steve. Two that Steve read—*Cosmic Consciousness* and *Anthropological View*—were especially significant to Steve. Dan believes that those books led Steve to believe that he was an enlightened individual. Back in California, Steve went to a yogi and said, "Help me. I am enlightened, and I don't know what to do!" Dan found that humorous. If a person is enlightened, Dan concluded, then he would know what to do. He wouldn't need a spiritual guide to direct him.

Dan's presentation that day revealed a fact that many of us had not known. Steve's adoption mom did not hold him for most of the first six months following his adoption. Steve's birth mother launched a lawsuit just days after his adoption because she changed her mind and decided to cancel the adoption. Steve's adoption mom feared she would get close to her new baby and then lose him if the lawsuit undid the adoption. As a result, she decided not to hold baby Steve to avoid developing close feelings for him until she knew that the adoption would hold. Dan revealed that, as a result of not being held, Steve suffered from what's called Russian orphan syndrome (also called reactive attachment disorder). Dan says he believes this had a big effect on Steve, though for a long time Dan was unaware of Steve's childhood and the impact it had on him. Dan found that Steve's past helped him understand Steve's complex personality and many of the actions he personally experienced.

Dan also shared that Steve was cruel in his earlier years. That was evidenced in the way he treated his girlfriend, Chrisann, and their daughter, Lisa. For example, Steve pushed Chrisann out of his life and denied that Lisa was his daughter. He even sued Chrisann and stated in his suit that she had slept with many other people and that, for medical reasons, he was incapable of being Lisa's father. When Apple released the Lisa computer, Steve again denied that Apple's Lisa computer was named after his daughter. He had Apple's PR department tell the media that Lisa was an acronym for "Local Integrated System Architecture." Apple's own software developers referred in-house to this so-called acronym as "Lisa: Invented Stupid Acronym." Computer industry insiders did not buy into the PR machine either and coined the term "Let's Invent Some Acronym" to fit the Lisa's name. Although Steve denied the computer was named for his real daughter, he later admitted that it was, but only after a court-ordered DNA test proved he was Lisa's father. Later in his life, when he interviewed with Walter Isaacson, he stated that "of course" Apple's Lisa computer was named after his daughter.

As another example of Steve's cruelty, Dan pointed out that Steve felt Dan had divulged confidential information about him in a much earlier interview. That one event was enough for Steve to shift Dan from a best friend to being totally blocked out of his life for nearly five years.

The beautiful thing about Dan's talk came in his closing comments. It was a healing moment. He said, "The Steve I knew in college, that I traveled around with, was kind." That is the Steve he said he would choose to remember.

Apple and Microsoft

Iron sharpens iron. So it was between Apple and Microsoft. A lot of innovation that people came to know on the Mac was pioneered by Microsoft. Simple things like double-clicking on the title bar of an application to zoom it to full screen first appeared in Microsoft Excel. Until that time, the user had to accurately position the mouse pointer inside a small square on the onscreen tool bar to achieve the zoom (or revert to keyboard commands or finding a zoom option in a menu.)

In my opinion, Apple served a critically important role in the success of Windows. The Macintosh was the first commercially successful window operating system. Windows 1.0 used a tiled window interface whereas the Mac used overlapping Windows. Some individuals within Microsoft referred to the Mac as "the best Windows demo machine."

The Microsoft-Apple relationship was competitive on many fronts and complementary on others. In 1997, Bill Gates and Steve Jobs negotiated Microsoft making a $150 million investment in Apple for non-voting shares. Apple had endured eighteen consecutive months of losses, making its situation dire. That Microsoft investment enabled Apple to stave off elimination and helped it rise again under Steve Jobs' second term at Apple. At the same time, Microsoft committed to a minimum of five years of development of Microsoft Office for the Mac. That commitment was arguably more valuable to Apple than the $150 million. It kept Apple useful and relevant in the business marketplace. Microsoft Office was the number one Apple application with over eight million units sold at that time. Even prior to this period, we were often reminded within Microsoft that, because of Microsoft Office on the Mac, we made more revenue per Mac than we made per PC.

Jobs and Gates

Much has been written about Steve Jobs and Bill Gates. Probably more about Steve than Bill, especially since Steve's death in the fall of 2011.[1] The last time I saw Steve Jobs in person and interacted with him was at the 2005 All Things Digital conference, also known as D5. He was as engaging and brilliant as ever. He definitely had a softer side than the Steve I had experienced years earlier in that hotel lobby.

When Steve and Bill spoke at the D5 conference, the room was divided. Some were Apple fans, while others were Microsoft fans. All stood in unison to applaud both at the conclusion of their discussion on stage. These two men have undoubtedly impacted our world with their unique approaches and visions more than any two people in the past century. D5 is an event I will never forget, as I will always remember the multiple opportunities I had over the years to interact directly with both Steve and Bill. I am grateful to both men. And I choose to remember Steve as a brilliant man whose gifts helped change our world.

---

[1] A book I would recommend on Apple and Steve Jobs is by Kelli Richards, *The Magic and Moxie of Apple: An Insider's View* (Cupertino, CA: The All Access Group, 2012).

SIX

The Power of Ownership and *COMMUNICATION*

Going Public

On March 13, 1986, Microsoft became a publicly traded company. Our stock went public at $21 per share. It rose to $28 per share by the end of the first trading day. The initial public offering (IPO) raised $61 million.

Bill Gates said he originally hoped that Microsoft stock options would help employees get a solid down payment for purchasing homes. Even with the vision he held and shared, he did not see the wave of millionaires, and even some billionaires, that would be created from the granting of Microsoft stock options.

The stock vested, meaning we did not have access to cash it out all at once. In fact, the first vesting did not occur until eighteen months after the grant. A portion of the stock would then vest each six months. Vesting a stock grant took a total of four-and-a-half years. Along the way, we had the ability to earn additional stock options, thereby starting the cycle again and overlapping any previous grants.

As a result of the vesting schedule, the stock options became known as "Golden Handcuffs." An exodus of talent often occurred right after the four-and-a-half-year mark as people completed a major vesting cycle. This became a real issue later in Microsoft's development as the stock price stopped its meteoric rise.

For the entire ten-year period that I was at Microsoft, the stock rose strongly. Along the way I heard many people share why they chose not to invest in Microsoft stock, believing it had already risen as high as it could go. Each incantation would be met with further rises in the stock price as it increasingly grew year after year. Frank Gaudette, Microsoft's CFO, helped convince those who believed the stock had reached its high that they were right. At the end of every quarter Frank would make public statements to the effect of "Yes, we had a good quarter, yet we are not sure that we can continue this growth at the same levels." The company's stock would often fall a little for a short time upon his pronouncement. Frank was a master at managing risk for the company.

Ownership matters. For a long time many employees in a vast number of companies used to take a job for life. They would stay their entire career at one company and get the gold watch, or its equivalent, at retirement. In today's marketplace, this is typically not the case. People in our day are expected to have six or more careers in their lifetime. Giving employees equity or profit sharing, as Microsoft did, can help you grow your business and retain the talent you need. We are all trying to do more with less, and retaining key talent is more important than ever.

- Are your employees vested in your business? It doesn't have to happen through stock options. Profit sharing is arguably as strong a reason, if not stronger, to convince employees to stay in many companies. Developing a culture that cares about its employees and demonstrates this by continued listening and investment in training are two additional ways to get your team to be vested in the business.

Communication Control

The Canadian Apple User Group Network

How do you win a war? In traditional war strategy, long-term victory often goes to the party that controls the communication channels.

In 1985, Lotus and Microsoft were at war. Lotus 1-2-3 was the dominant spreadsheet on PCs. With the release of the Macintosh, however, the new platform was up for grabs. So Lotus announced Jazz. Jazz was to become Lotus' attempt to continue its spreadsheet dominance on the new Macintosh.

In 1985 and 1986, user groups were the dominant way that those who were interested in technology learned of emerging products, met with other interested technologists, and shared their insights. Every major Canadian city had one or more user-group meetings. These groups typically met every month. Some met even more frequently. User groups were stand-alone entities. Each group had its own leadership, agenda, and approach.

As part of the strategy to win the technology war, we (Microsoft Canada) founded the Canadian Apple User Group Network and connected all the individual user groups. Our strategy was to offer communications across all groups, share templates for the new Excel platform, promote best ideas from one group to the other, and do so via a published newsletter, attending meetings as often as possible, and sharing sneak peeks of emerging technology. By being the publisher of the news that

Kevin Jampole (left) receives User Group award from Dave Jaworski (center) with Malcolm MacTaggart, Microsoft Canada

was sent to all the groups, we controlled the communication channel. We featured Microsoft products as well as products from all other major software developers, with the exception of Lotus. When Lotus won a packaging award for Jazz, we gave that a nod in the newsletter. We also gave awards to the best technology innovators in the various groups and featured them in our newsletters. The strategy worked. Microsoft Excel never let Lotus Jazz gain a foothold in Canada.

- What communication strategies can benefit you in your industry? Can you bring organization and information to add value to your channels, and at the same time, give yourself a strategic edge? Can you strategically control the communication channels?

The Importance of a Shared Vocabulary (MAG)

"Does anybody have an elastic?" I asked.

Heads popped out of office doorways in our Silicon Valley office.

"A what?"

"He wants a rubber band" came the response from another doorway.

My Canadian heritage gave me away again as a foreigner. Even simple things had new terms.

In Canada, a sofa is known as a "chesterfield." A "napkin" is a feminine hygiene product. The thing you'd use to wipe your mouth is a "serviette."

After countless jokes about my Canadian-isms, I decided it was time to celebrate our differences. Or, as they say in Quebec, "Viva la difference!" A date for a "Canadian Beach Party" was set for the Microsoft west region. It would include Canadian beers, such as Molson Canadian and Labatt's Blue, and various confections, such as Smarties (not the same as the US candy of the same name), Eatmores, Macintosh Toffee, and more. The entertainment would feature all Canadian bands, such as the Guess Who and Bachman Turner Overdrive, both from my hometown of Winnipeg. And there would be touques for everyone, including the kids.

As the date for the party drew closer, I noticed a lot of whispering and discomfort in the office. I called a few fellow employees together to ask what the rumbling was about.

"It's about drugs at the party," one person said.

"Drugs? Who is bringing drugs to the party?" I asked.

"You said you are" was the matter-of-fact reply.

I was stunned. "What? When did I say anything about drugs?"

"You said there would be tokes for everyone."

"It's toques, pronounced 'tooks' (rhymes with 'Lukes')," I clarified. "*Toques* is a Canadian word for a hat."

The concern was alleviated. Those who wondered if they should skip the party decided it was okay to come. The Canadian snowboard team had established a reputation for all Canadians as being tokers during the recent Olympic Games, and my Microsoft team apparently thought I was on that team.

Here are a few pictures from the Canadian beach party. Notice their touques.

Left to right, Diane Connelly and my wife, Susan Jaworski.

Our Microsoft kids

The lesson driven home in that silly example would be repeated at Microsoft many times. Except it was not always about Canadian-isms.

When a new person would join our team, it wouldn't take long before that person would be overcome with a puzzled look and say something like, "I have no idea what you all are saying." New employees often felt as if they had landed in a foreign country where everyone spoke a different language.

"Oh, you don't have MAG?" would be our response.

"What's MAG?"

"The Microsoft Acronym Guide."

We had developed a shorthand vocabulary that was natural to us. We didn't even realize how deeply we had slipped into it until someone new questioned what we were saying. New team members were encouraged to speak up any time they caught us speaking in confusing terminology. The worst situation was when some members were too embarrassed to say they didn't understand us. If they spoke up at all, the conversation would be a few miles down the road before they would finally admit that they were totally lost.

Confusing terms made their way into our software too. Today most of us are familiar with clicking on a "File" menu and then choosing options, such as "Open," to access a file. "File – Open" was originally "Transfer – Load." In testing our applications with customers, we learned that simply changing a few words in a menu could change a task from confusing to simple and easy.

TAKE ACTION!

- Do you have an insider's language that you speak? Does it sometimes confuse your own team members? Worse yet, does it confuse your customers?

SEVEN

Creativity, Problem-Solving, AND INTEGRITY

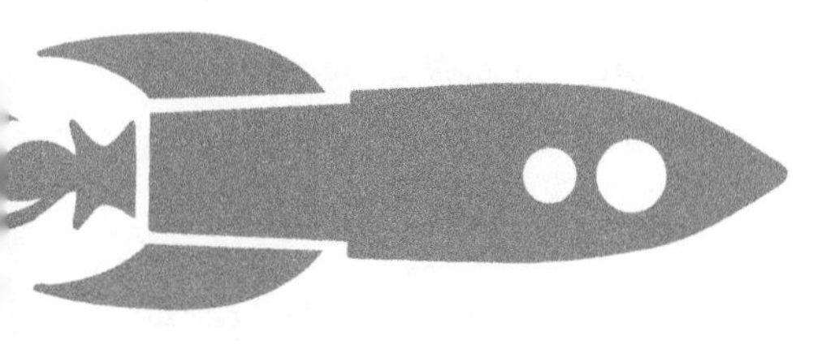

The Creation of Microsoft Office

Businesses write business plans. People write life plans. As the joke goes, if you want to make God laugh, tell him your plans.

One of the key skills in any business is problem-solving. If you want to build a creative culture that solves problems, you'd best be a leader who asks questions and who admits you don't have all the answers. Stay curious. Be open to the solutions to your issues coming from unexpected places. Here is just one example of this from my experience at Microsoft.

In 1986, Microsoft File was released for the Macintosh. We had pallets of this product. By the beginning of 1987, Microsoft Word and Microsoft Excel were selling very well. But Microsoft File—not so much. File was a basic database program. Most people seemed happy to use Excel or other products for such tasks, but not File.

In order to turn the situation around, our general manager, Rich MacIntosh, called for a problem-solving strategy session. Six of us sat at a table and brainstormed possible solutions. I believe it was Hans Taal, our warehouse manager, who first suggested shrink-wrapping File together with Excel and Word. Hans did just that. He took the three product boxes for Word, Excel, and File and shrink-wrapped them together. We called the bundle Microsoft Office. We touted the common-user interface that existed across all three products. The bundled products sold.

Hans then proceeded to have packaging made that looked the same as Word, Excel, and File. He removed the manuals and discs from all three software packages and loaded them into the new Microsoft Office boxes. This bundled version sold even better.

Not long after that, we called down to our US operations and suggested they try the same thing. Laura Jennings led the effort to bring Microsoft Office to the USA in 1988. And the rest, as they say, is history.

On February 19, 2015, I was invited by Microsoft Alumni's CEO Rich Kaplan to join a group of fifteen alumni at the Microsoft campus in Redmond, Washington, at the Executive Briefing Center. During that day we received presentations from Windows 10 product management, Microsoft Azure, and Microsoft Ventures, toured Microsoft's "Vision for the Future," saw Microsoft Security, and met with Microsoft Office product management. We discussed the origination of Microsoft Office and how far it had come. I shared what I thought then was the most recent number of Office users: over 800 million. I asked if that was the correct number. The product manager for Microsoft Office stated that, as of that day, there were more than 1.3 billion active registered users of Microsoft Office. Our little brainstorming session in 1987 had paid off beyond our wildest dreams!

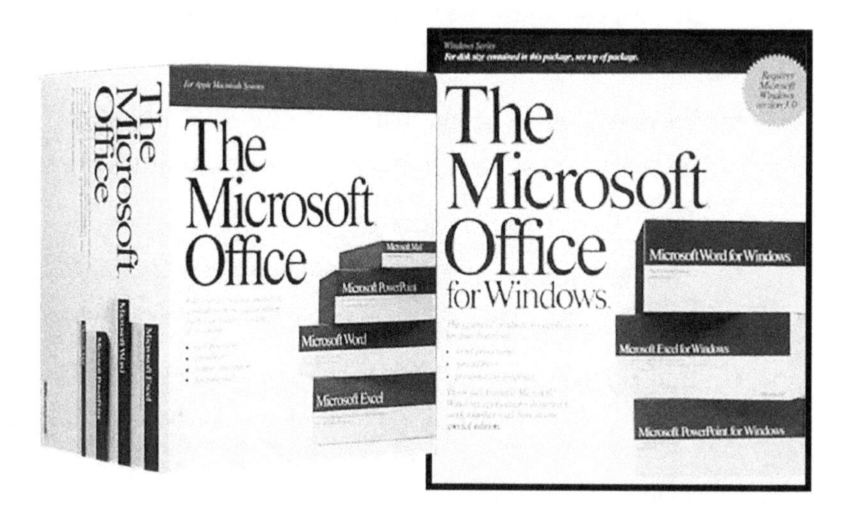

Microsoft Office for Macintosh 1989 (USA) (left) and Windows 1989 (USA) (right)

Even though the creation of Office started as a problem-solving exercise for Microsoft File sales, Microsoft File did not make the longer journey. File was eventually replaced by Microsoft Mail in the Mac version in the first US versions of Office and was never made into a PC product.

SmartPages

During the Microsoft years in Los Angeles another problem-solving opportunity eventually gave rise to a solution for all of our business and technology partners.

Each and every sales person in the field received sales presentations from Corporate, updated strategic product information, marketing programs and policies, company directories, pricing information, HR documents, and much more. The challenges were numerous:

- The paper-and-disk flow made every sales person a filing clerk administrator.
- Their skills as an administrator determined if they were presenting current and accurate information to their customers or not.
- Maximizing time spent with customers versus pushing paper and electronic files was important for any sales team.
- And the typical laptop's hard drives of the time period would easily get filled with all of the information being sent.

The solution? SmartPages.

SmartPages was first created by Dave Staehlin and I to solve these and other issues and to regain all that administrative time back for selling. SmartPages simply had one person gather all the latest information and then publish one CD-ROM disc with all the current information to every field sales person. As long as you had the current CD, you knew you were looking at all the current information. Each month you'd simply take out the old CD and insert the new one, and then all your filing for this critical information was done.

SmartPages helped accomplish another goal. Most computers didn't come with CD-ROM drives. Laura Jennings, Microsoft Office product manager for the Macintosh, worked with us and Toshiba to get every sales person an external CD drive. Every sales person was now armed with this powerful new technology and was informed with hands-on experience, making them more knowledgeable and valuable to their clients.

Rob Glaser, future CEO and founder of Real Networks, was a product manager on a CD encyclopedia called Encarta. He called SmartPages the first real use of the new CD-ROM technology for business that he had seen.

SmartPages became an indispensable tool for our field team. Soon we were getting requests for the discs from other departments all over Microsoft.

SmartPages also birthed other similar uses of the technology. Microsoft created discs for customers and developers. These discs contained current versions of documentation and other relevant information for the specific audiences.

In time the computer networks became fast enough to let us setup private shared file folders that contained the information. You no longer had to replace the disc or worry about the information you had being current. Updates were now made in real time, as long as you were using the file from the network.

- Are there ways you can automate or even semi-automate parts of your business to gain leverage and keep the focus of you and your people's time on the actions that move the needle? Can you do the same for your customers and business partners?

Integrity Is a Difference-Maker: The Benefit of Being a Giver Versus a Taker

Microsoft released Access in 1986. Today many know Access as a database product, but in 1986 that was not this product. The first time Microsoft used the name "Access," it was used for a communications program.

Internally the Microsoft marketing team boasted that Access could host up to eight simultaneous communication sessions over modems. That meant the market for it was primarily financial traders and news related organizations. To most others, this ability didn't matter. Few had any practical use for more than one modem session, if they indeed knew why one was useful.

Frantek, based in Ottawa, was one of Canada's main computer technology distributors. As Frantek prepared their initial order for Access, they suggested they were going to purchase two hundred units. Believing they were targeting the general consumer population and not the financial traders and news organizations who might value such a device, I suggested they buy ten units and see how sales went. I did not believe the market demand existed for Access as envisioned by our Corporate marketing team. I do not believe they even sold those ten. Although it meant taking a smaller order at the time, doing the right thing is always the right thing for the long run. As a result of interactions like this, Frantek called me to ask my opinion on third-party products. Frantek's executives trusted me. I gave them my honest feedback, good or bad. Our relationship grew.

In November 1987, Microsoft released Excel for the PC platform. Frantek said they were going to order twenty units to sell to their dealers. I suggested they order two hundred, which is what they did. All those units sold.

At the time, the software industry pretty much worked this way: software manufacturers sold to distributors, distributors sold to dealers (retail sales outlets), and dealers sold to businesses and consumers. The largest dealers like Computerland and Egghead Software bought direct from the large software manufacturers. Software manufacturers paid rebates and marketing incentives to the distributors and dealers based on their purchases. This resulted in distributors and dealers often placing large orders at the end of a three-month quarter period to get the marketing bonuses. The bonuses were paid on these purchases, also known as "buy in," by the retail channel.

Rich MacIntosh, general manager of Microsoft Canada, later took even further the concept of serving our distributors and dealers with an aligned business objective. In doing so, he innovated one of the largest shifts in retail channel behavior. He gave up control and changed the way the industry sold and marketed software in the process. He eliminated buy-in incentives and made all marketing bonus payments based on sell-through. In order to get specific bonuses, the product needed to sell all the way through the channel to a business or end user. If it was sitting on a distributor or reseller's shelf, we were to think of it as still unsold. The sale was only final when a business or customer purchased it.

- Are you honest with your customers? Would you tell them to order less if you thought that was in their best interest? Does your team know that you want them to take the long-term stance for your customers? Do you have the same honest candor internally, positive or negative, for the long-term benefit of your business?

People Are Our Greatest Assets

Relationships Matter

Microsoft is identified as a technology company. Yet, first and foremost, it is a people company. And whether other companies realize it or not, they too are about people—or at least they should be. Companies are about relationships: relationships with each other, with customers, and with other people within each company's identified industry. These relationships make or break opportunities for every company to grow personally and professionally.

Our personal lives are all about relationships as well. We are relational beings. That is how we have been created. The Bible says we have been created in God's image (Genesis 1:26–27). As a Christian, I believe in the holy Trinity: God as the Father, the Son, and the Holy Spirit. The Trinity shows that even God is in relationship with himself. So it is a natural fact about us that we are built to be in relationship with each other and that this is central to our journey as humans, including for the organizations we create.

Real Heroes Behind the Scenes

Every story has its unsung heroes—those people behind the scenes who influence outcomes in the most significant ways. Microsoft was no different. Here are a few of the people I would single out as Microsoft's unsung heroes.

Rich MacIntosh

A friend of mine had a doctor named Dr. Payne and his dentist was named Dr. Hurt. I know a lawyer whose name is William High, and he goes by Bill. (Lawyers are known for high hourly billing rates.) It's ironic how some names fit the chosen career. The naming irony holds true for Rich MacIntosh. He would go on to be one of the important personal computer industry influencers for both IBM-compatible PCs and Macintosh computers. We are all richer for this.

Rich MacIntosh had been a leading executive in the Canadian computer industry. He was selected as the first employee of Microsoft Canada. He was a leader in every sense of the word, including in mentoring others. Everyone should be blessed to have a mentor like Rich. Many times I carpooled with him into Microsoft Canada. On these rides, in meetings, and in many other shared experiences over the years, he coached me on personal as well as professional matters. I truly felt he took a personal interest in me, my family, and my career. He did that for many others too. As I look back on those car rides, I see the wisdom of Rich in mentoring me without my even recognizing it at the time.

Rich was a true pioneer in technology. The contributions he made to Microsoft helped build its success on many fronts. I will highlight just a few.

As shared in the last chapter, Rich changed the software marketing incentive process from buy-in to sell-through. He correctly determined that the only thing that mattered was getting software into the hands of an end user. Until that happened, the sale was not truly a sale.

Most distributors and resellers had return privileges, and those return privileges meant that, until an end user had purchased the product, it could be returned. To cut down on returns and to get the software to the end user, Rich implemented sell-through incentives. Now everyone in the process—the software company and its sales people, the distributor, and the reseller—would receive incentives based on sell-through. Everyone's goals were now aligned. This meant that sales representatives would not receive their incentives until the product had been reported as sold all the way through the channel.

When I joined US Sales Operations at our head office in Redmond, Washington, I discovered that it was taking more than a year for that indirect sale to get tracked and reported from end to end. As a result, commission checks that came to sales representatives were more like Christmas presents. Sales representatives did not see a direct correlation of their actions to the commission checks they received. That was definitely not an ideal situation, and I was tasked with reducing that time differential. Our team and our channel partners worked together until we got the process down to thirty-seven days. Now that check, arriving in the hands of a sales representative, had a clear link to recent sales.

Rich changed the way the entire industry worked. He had established the way needed to align the end goals of all involved. It took some other adjustments to fully see the benefits of this huge change, but it was Rich's vision and commitment to what really mattered that paved the way. Now the focus was where it should have always been—on the customer who had purchased our software for his or her use, versus the middle steps towards that end goal. This shift changed the transparency of the entire lifecycle of a product's movement from creation to use. The entire industry eventually followed Rich's lead.

Here's another way Rich changed our industry. Our corporate customers were initially only buying fifty to a hundred products. As computers became more prolific in organizations, that number grew to thousands. And many of those computers had a variety of products on them, such as word processing, spreadsheets, and databases. Each one of those products had a box with a manual, discs, and the most important ingredient, the software license. As personal computers started to get connected to each other via office networks and documentation moved online, only a record of the software license was needed. Yet software manufacturers were still requiring companies to buy and warehouse a box of software for each and every computer they operated. Many corporate accounts showed our team rooms of boxes where they kept software to show that they were properly licensed. The amount of space required was becoming increasingly ridiculous. Understanding the problem, Rich changed licensing to a paper agreement and eliminated the need for the storage of boxes. It was a philosophical shift to recognize the value of the software as a digital entity that did not require a physical presence. Again, the industry followed Rich's lead.

- Are your goals aligned with your sales channel? Are your incentives aligned with the behaviors and actions you deem necessary to achieve your defined goals and objectives? If not, why?

- Is there any way you can improve your relationship with your customers by making their life easier with your product or service?

Rich MacIntosh and Scott Oki

Rich MacIntosh and Scott Oki were like Batman and Robin or the Lone Ranger and Tonto. They were a dynamic duo who helped us understand the benefits of working in teams.

Scott Oki had come from Hewlett-Packard. He boldly projected that the international market would one day represent 50 percent of Microsoft's business. Then he set about to turn that projection into a reality. International sales grew to exceed 50 percent of Microsoft's revenue. Software turned global.

True Pioneers: Scott Oki (left) and Rich MacIntosh (right)

Scott presided over all of Microsoft's international business. Rich led Microsoft Canada to become one of the fastest growing subsidiaries in Microsoft's international portfolio. Meanwhile, back in the US, executive management was not pleased with its US Sales and Marketing leadership. Rich and Scott were tapped to come to company headquarters in Redmond, Washington, and lead Microsoft's Sales and Marketing efforts in the US.

The combination of Rich and Scott created a wave of success that drove Microsoft from dead last in almost every product category to first place. They did this by studying Microsoft, its competitors, and even other industries. They strove to see what best practices could work in the newly emerging technology industry.

As part of this endeavor, Scott had us reverse engineer the profit-and-loss statements of major competitors. If the company was public, he would use a combination of the information available, plus any statements made by the executives, to understand their strengths and weaknesses. If the company was private, he relied on any information we could gather. I remember Scott telling us that if we were successful, the CEO of Lotus would make certain statements and these would indicate we were achieving our objective. Jim Manzi, Lotus' CEO, later made those statements almost verbatim. We instantly recalled Scott's words. We knew we had hit the bull's-eye with our strategies and that we had pierced Lotus' armor.

Bob O'Rear

Although Bob O'Rear appears in the picture of the original eleven company employees of Microsoft, he is known by few outside of the company.

Bob O'Rear (top) and Bill Gates (bottom)—then and now

Prior to joining Microsoft, Bob O'Rear worked for NASA. He led the calculations that brought America's astronauts safely back to earth after going to the moon and leaving earth's atmosphere. A mistake in his math as these adventurers flew off into space would have sent them way off course, perhaps never to be seen again. A mistake in his calculations for the astronauts' re-entry angle to earth could have led to their burning up on their way down. There was no room for error in his math. I had heard that Bob had done the calculations on a slide ruler. That proved to be a myth. Yet when Bob corrected the myth, he let me know how much more complex his work was than simply calculating an angle for exit and re-entry.

> In my work with NASA, I never used a slide rule. The re-entry program I worked on was mostly programmed in Fortran and ran on a mainframe (Univac 1110). The answers it produced were transmitted to the command module and were a firing sequence of the small jets that lined the command module to cause it to carve a path into the atmosphere and rotate to dissipate heat. I did use some of the real early calculators to formulate or find mistakes in the algorithms used for these calculations—but not a slide rule.

What Bob did, given the limited knowledge we had at the time and the grave implications for the slightest error, was truly amazing.

So how smart was Bob? And how key was his work to Microsoft's edge in technology over other emerging tech companies who were vying for early leadership? When Intel was releasing new chips and Microsoft needed to write its operating systems software on them, Bob would write an emulator in software on an existing machine based on the written specifications Intel provided. He did this before the new chip existed or was available so that Microsoft engineers could develop software taking advantage of the new technology as early as possible. His innovations gave Microsoft a competitive edge.

As Microsoft expanded operations globally, Bob would travel to new territories and learn what was required to setup new subsidiaries, including the local business requirements. He would also hire the general manager and initial staff members and support them in getting the operation up and running. He did this over and over, like Johnny Appleseed did in planting apple trees worldwide. The fruits of his seeding made Microsoft a global powerhouse.

Jon Shirley

Jon Shirley and Bill Gates

One sign of Bill Gates' brilliance was how he surrounded himself with other brilliant thinkers. Bill had his own dynamic duo team of himself and Jon Shirley. Jon was the president and chief operating officer of Microsoft as well as a member of the Board of Directors.

Jon Shirley came from Tandy Corporation. He had a brilliant analytical mind, and he understood the value of sales. He masterminded with Bill so that each subsidiary was held accountable to key measurable objectives. You could hand him a stack of spreadsheet printouts, and in a few seconds he could find an error buried a few pages back where things just weren't adding up. He listened with intense focus as you defined the strategy for your area of the business. He asked penetrating questions. He got to the heart of the key success factors almost instantly. If Jon didn't ask the tough questions, Bill did. Between the two of them, it seemed nothing was missed.

Given how good Jon and Bill were together, preparing for a business review was an intense exercise in itself. We would think through the main points that were to be made and the data that would support those points. We also tried to think of every question Jon and Bill could ask, along with the answers to those questions. We would prepare as many slides, sometimes more slides, than the main presentation to be ready for the questions that could come. The quality of the interactions with Jon and Bill led us all to be better. The intensity of our preparation helped us think through our own proposals and strategies at a deep level. We were all better for the experience.

Jon served for seven years as Microsoft's president. I have no doubt that he would have continued much longer if it hadn't been for a double perforated bleeding ulcer. It could have killed him, but thankfully it did not. As a result, Jon stepped out of day-to-day management and became an important member of our Board of Directors.

Frank Gaudette

Frank Gaudette was Microsoft's chief financial officer (CFO). He was also Microsoft's first head of HR, Facilities, and Operations. He came to Microsoft from Frito Lay. As important as the numbers were and are, Frank was probably most known for his statement that "People are our greatest assets." He meant what he said, and he lived it.

When Susan and I were offered the opportunity to move to Los Angeles to work for Microsoft there, Frank reached out to see if we needed any assistance. We were young and did not have the asset base to support moving from the countryside just outside of Toronto to the real estate markets of Los Angeles. Frank recommended and provided financial assistance from the company to make that happen. He shared his belief in me on many occasions, something he did with many others too.

To Wall Street, Frank was the guy who continually stated, "Yes, we had a great quarter. It is uncertain if it can be done again." This competent approach achieved two objectives: (1) it kept the stock growing at a steady and reasonable rate, avoiding a roller-coaster of ups and downs, and (2) it avoided shareholder lawsuits for over-promising future results in the event that we experienced a down quarter. Frank was a confident and competent financial and operations leader. And regardless of the stress that came naturally from managing Wall Street expectations and building out Microsoft facilities to keep up with the rapid growth of the company, he was filled with joy. He exuded joy and laughter while at the same time exhibiting a hardcore nature about all things financial.

In 1992, Frank came down with a fast-growing Grade B lymphoma, which is cancer of the lymphatic system. He died almost nine months later in April 1993. Microsoft and the world lost him too early. He was a great man who truly understood that Microsoft's great technology was the derivative of great people and a great work environment.

Mike Maples

In my opinion, Mike Maples is the best software development manager ever. Mike came to Microsoft from a successful career at IBM. Many people wondered if one could make the leap from IBM's culture to Microsoft's in such a high-level leadership position. Mike, however, made it look effortless.

From left to right: Mike Maples, Steve Ballmer, and Bill Gates

Under Mike's leadership, Microsoft shipped more major products in a short eighteen-month window than any other major software development company in history. Mike brought the goal of "zero defects" to Microsoft software development. The quality that he brought to all Microsoft products cemented our position as a producer of high quality software that companies could rely on.

While there were tens of thousands of people who contributed to Microsoft's success, the leadership secret sauce behind Microsoft's rise from worst to first was Bill G. plus Jon plus Frank plus Mike plus Rich plus Scott plus Bob. Add in Brad Silverberg, Pete Higgins, Lewis Levin, Charles Stevens, Chris Peters, Ron Davis, and many other talented team members, and one can see how Microsoft became unstoppable. Together, we went from worst to first in every major category we targeted.

The Chairman's Award

Microsoft International held its annual meeting in Cannes in the south of France in 1986. Microsoft executives decided to create a new award, which they titled the Chairman's Award for Excellence. Microsoft's Bill Gates, chairman and CEO, would present the award. Executives across the company were asked to nominate their top people for this special recognition. The executives then chose one winner from all the nominees. In presenting the award, they stated that people from all positions in the company, including sales, finance, and marketing, were eligible for it. I was blessed to receive the first-ever Bill Gates Chairman's Award for Excellence.

Chairman's Award 1986

In 1987, the Chairman's Award was given to an individual in international and a second Chairman's Award was given to an individual in the United States. As the company grew, more Chairman's Awards were given each year.

Dhiren Fonseca, who I recruited to Microsoft after having met him at the Macintosh user-group meeting several years earlier in Winnipeg, Canada, was also a recipient of the Chairman's Award. Dhiren joined me at Arabesque Software for the release of Ecco in 1993. A year later the company was sold, and Dhiren returned to Microsoft. Dhiren became a founding executive of Expedia. He retired from Expedia in 2014 after leading them to become a dominant force in the worldwide travel industry.

From left to right: Doris Jackson, Tammy Roark, Dhiren Fonseca, and Celia Paget

People. Relationships. Creativity. Commitment. Integrity. Problem-solving, especially in teams. These were some of the most important elements working together to move Microsoft forward, including in those formative years of 1985 to 1995. They are still essential elements for any company that wants to move ahead and become a leader in their industry.

E I G H T

Wisdom of a TWO-YEAR-OLD

The most important relationships are at home.

When I joined Microsoft, Susan and I had been married for a little over two-and-a-half years. We had a one-year-old daughter, Jennifer, and a second child on the way.

Amanda was born to us on September 1, 1985. Susan and Amanda came home from the hospital two days later. "My parents will arrive tomorrow," I said. I thought that would be all the support she needed, so I grabbed my luggage and left for the airport.

Susan reminded me that I never told her about the part where we'd leave our hometown of Winnipeg. Susan was seven months pregnant when she made the move from Winnipeg to Mono Mills. In the early days of our new life in the suburbs of Toronto, I travelled extensively, leaving her alone in a new city and with no friends. It put a strain on our relationship. At the time I missed understanding the impact this had. Rich told me that, as hard as I worked, there was no way I would be able to stay married. I told him he was wrong, but he turned out to be almost right. Those early years took quite a toll on my marriage.

My wife Susan, daughter Jennifer, and I

One evening Susan, Jennifer, and I were sitting at the dinner table. Jennifer was all of two-years-old. I said to Susan, "I wonder if Jennifer understands what the word Microsoft means since I use it in conversation quite often?"

Without missing a beat, Jennifer replied, "Yes Daddy. That's where you live."

Susan looked at me with eyes that said, I didn't teach her that. Jennifer is saying that on her own.

"No, Jennifer," I replied. "That's where Daddy works."

"No, Daddy," Jennifer said. "That's where you live."

I felt like I'd been skewered with a Samurai sword. Guilty as charged.

While Susan and I were surprised with Jennifer's candid response and certainty, she maintained a somber look throughout the exchange. To her, she had simply stated a fact and didn't understand why we reacted as we did.

I had been blessed with winning many awards and getting strong performance reviews, stock options, and promotions. Yet there was an unfortunate side to this. The strain on Susan and my relationship with her had grown. As I traveled more, up to 85 percent of the time, she had less desire for intimacy in our relationship. I was winning at work and getting less satisfaction at home. Rather than turn more of my attention to Susan, Jennifer, and Amanda, I turned to my job and worked even harder. I admit I had natural workaholic tendencies to begin with. Of course, my response only exacerbated the situation at home. The situation had become so bad that when Susan became pregnant with our daughter Sarah, she denied it. She said there was no way she could be pregnant because I had not been home enough.

When the pregnancy was announced, the team at the office joined in the disbelief and made comments like "Are you sure it is your child?" Sarah was later referred to as "the Miracle Child" for these same reasons. It seemed everyone was aware of how extreme the situation had become—everyone, that is, except me.

Dennis Rainey is CEO of Family Life, an organization that helps people find biblical guidance for marriage and family relationships. Years later, when I had the opportunity to meet with him, I described the challenges that continued in my marriage. Dennis asked if I had ever had an affair. That was easy. "No," I honestly replied.

He went on, "With work?"

An affair with my work? That took me by surprise. I had never thought of my workaholic tendencies as an affair in my relationship with Susan. I wanted nothing more than the best possible relationship with her, and yet I had been sabotaging my marriage. I had been putting almost all my energy into Microsoft and funneling little into my family. I found it easy to "justify" putting a safety border around myself, holding back what I saw as a little to avoid getting hurt by Susan's rejection. I estimated this as a 5 percent holdback.

Dennis said, "Don't be surprised if Susan would feel it more as 50 percent."

Around that same time, my friend Steve Johnson, a former pastor and now a leader of the mentoring ministry 2X Global, talked to me about priority. On our wedding day, the commitment we make is you, my spouse, are my priority. He shared how, when we first get married, the word *priority* doesn't mean a lot. In the early years, less competes with our marital relationship, so seeing it as a priority is easier. The word *priority* only truly takes on meaning when we must make hard decisions about where we put our time and energy. And that typically happens as our careers take off, our families grow, and many more people and activities compete for our time.

Susan and I stayed together—and God helped make this happen. I am thankful to Him for keeping us together and helping us work toward strengthening our relationship. The process required candid discussion, years of prayer, and a never-give-up attitude on both our parts. As of this writing, Susan and I have celebrated thirty-four anniversaries. God is good.

- Do you have a handle on your true priorities in both your personal and professional life? What are those priorities, and how are you meeting them?

- Many people drop everything to go after a loved one when their relationship falls apart. The right question to ask yourself is, why wait until that moment? Start now, before it is too late. In this regard, I recommend Andy Stanley's book *Choosing to Cheat: Who Wins When Family and Work Collide?* Stanley offers essential help and hope to anyone willing to change.

Doug and the Blibbets

What's a blibbet? The blibbet was the stylized "O" in the original Microsoft logo. We formed a musical group within the company and decided to carry on the blibbet in the band name since it had been dropped from the new forward-leaning Microsoft logo with the Pac Man bite out of it. Microsoft had decided that it was time for a change. Our group decided we had now found our brand.

We formed the group Doug and the Blibbets to perform at Microsoft's National Sales Meeting (NSM) in 1989 in Seattle. We then did an encore performance in 1990 at the NSM in Hawaii. The band members were Doug Martin on lead vocals, Diane Connelly handling vocals and percussion, Joe Vetter playing keyboards and on vocals, Jim Vetter handling guitar and vocals, Marty Taucher playing the drums, and me on guitar and vocals. Mark Dickison, CEO of Trifilm and the creative genius behind all the inside "spoof" videos created for Microsoft meetings, was our manager. We practiced for months and worked up thirty-four songs for a full night of music. We only performed twice, and it was worth it. Playing was our joy. And everyone at the NSM had a great time.

NSM Hawaii—break time between sets

The Hawaii event turned into an expensive event. My butcher block Fender Stratocaster pictured here was stolen. Or perhaps it simply loved Hawaii too much to leave. Either way, this costly instrument never made the trip home.

Truth be told, I had another agenda for forming and playing in the band. If I was a band member, I wouldn't be dancing with other women. I thought that would make Susan feel better about the NSM. As things turned out, that didn't matter. She still hated the thought of all of us dressed in tuxes and having all-night parties with no spouses allowed anywhere near the place. I understood how she felt about this. The fact that spouses were told not to come within fifty miles of the event "or you risk being fired" was a horrible message to send to those who supported us all year. We worked hard. The victories Microsoft achieved were family victories, not individual ones or even just corporate ones. Spouses played a huge role in supporting their mates through the long workweeks, which often included weekends. They should have been invited to celebrate the company's success with us.

To add salt to the wound, Microsoft sent the promotional information about the upcoming NSM to our home addresses and started doing so months in advance of the event. That just meant many months of frustrated spouses.

I recently received confidential access to an internal study completed by a major US corporation with over eighty thousand employees. Employee wellness was assessed. The study found that the number one stress at work for their employees was stress in the relationships at home. If things were good on the home front, it was easy for the employee to be focused, engaged, and productive at work. If things were stressful and challenging on the home front, everything became a challenge at work.

- Are those who matter in your life treated like they are true priorities? If not, why?
- Is your business friendly to the significant others who support your team members doing great work?
- What is the number one stress to your team? What can be done to help alleviate that stress?

NINE

No Risk, *NO REWARD*

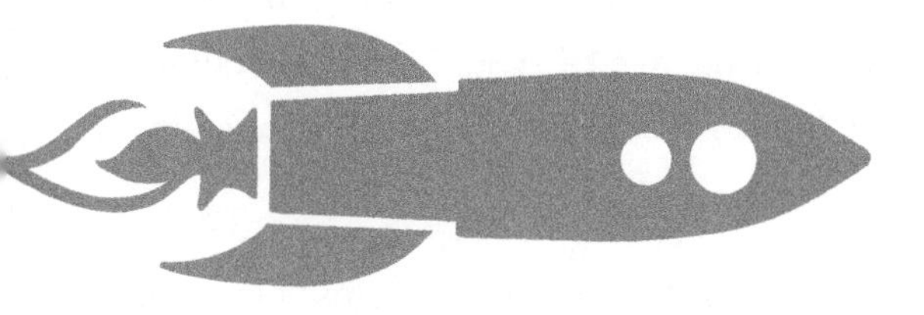

Scott Oki and Rich MacIntosh made the transition from Microsoft International to run US Sales and Marketing. During this timeframe, Rich called me aside: "I am not going to promote you to lead Microsoft Canada as I want to figure out where you are most needed in the US and then bring you down to help us."

Malcolm MacTaggart, who led Microsoft Marketing, was promoted to general manager of Microsoft Canada. Shortly thereafter, the phone call came as Rich said it would. "We want you to move to Los Angeles and run Western US," said Rich.

At that time, Microsoft divided the US into three regions—West, Central, and East. Each region was responsible for resellers and distributors located in the geographic area, as well as all corporate, educational, and state and local government business. Ed Johnson was GM of Central US and Mike Appe was GM of Eastern US. I was asked to lead Western US. Paul Burden and his team led all federal government business in Washington, DC.

By fluke of geography, I ended up with 80 percent of Microsoft's US reseller and distributor budget. You see, the head offices of our major distributors—Ingram Micro D and Softsel/Merisel, as well as the headquarters of Egghead Software, Businessland, Computerland, and others—were located on the West Coast. The West Region also developed and serviced Microsoft's largest corporate account in the world—Boeing.

Boeing was located in the Seattle area, right in Microsoft Corporate's backyard. And that was the problem. Boeing standards were Lotus 1-2-3 for spreadsheets and WordPerfect for word processing. They purchased roughly $100,000 of Microsoft product. That was a small number for the size of Boeing. They represented great potential for us.

We did not have a sales office in the Seattle area. The Pacific Northwest corporate sales were serviced out of the San Francisco office. We learned that both Lotus and WordPerfect were planning to open offices in our Corporate headquarter's backyard.

Product managers from Microsoft headquarters would come and go as they liked at Boeing, viewing them as an account with great potential and a local corporate account-testing lab. Little did the product managers know that Boeing saw this "come and go as you like" attitude as disrespectful.

Lloyd Wilhelms, who I had worked with at Citation Software, was recruited to join the Western region team. Lloyd had gone on to work with Motorola and had significant success in corporate sales. He established a strong presence for us in the Northwest for all our business. Lloyd recruited a strong team. On one of his first meetings with Boeing, Lloyd met with twelve Boeing computer services managers. They informed him that they were close to barring Microsoft from coming into their facilities due to the continual uninvited visits from Corporate team members. Lloyd assured them we would get things under control. He listened well and put a plan in place to have his team demonstrate products of interest to Boeing teams. Lloyd's team zeroed in on the opportunity with Boeing teams working on the 777 program. They were already using Macintosh computers for some of the work. Lloyd, his systems engineer Tom Davis, and his demo expert Cynthia Scott, won a major deal for Microsoft software. Lloyd and his team continued to develop wins within Boeing. Eventually he brought on Carolyn Hathaway to manage and grow the Boeing account for Microsoft. She came from Digital Equipment Corporation (known in the industry as DEC) where she had worked on the Boeing account. As a result of Lloyd and his team's careful listening, creating interest, demonstrating products, and providing excellent systems engineer support, Boeing stayed with Microsoft. They became Microsoft's fastest growing corporate account in the world and our largest account. Microsoft employees went from being warned about getting kicked out of Boeing to being invited by Boeing to their strategic planning sessions.

Lloyd's success came with a cost. He and his team told product and project managers that they needed to work through them to work with Boeing. Many at Microsoft thought this was a power play, but it was not. The risks were real. Microsoft could have lost Boeing, and Lloyd and his team knew this. Though their efforts didn't make them the favorites with some at Microsoft, they managed the Boeing account with perfection in those trying times.

Lloyd and his team invited Boeing to Microsoft Corporate. Taking the Boeing senior IT managers for an executive briefing with both Microsoft leadership and product teams was a win-win for both companies. Boeing saw and had input into our product strategy; and the product teams received customer input directly from decision-makers. These high-level executive invitations to Microsoft were repeated and became the genesis for the Executive Briefing program, one of Microsoft's major marketing programs that continues to this day.

Boeing helped us understand the need for a major change in the way we worked with corporate accounts. Microsoft and other companies in the software industry would do major product releases approximately every eighteen months. Corporations couldn't accurately budget for the release cycles when they varied from product to product and release to release. Working with Rich Macintosh, Lloyd's team pioneered the Microsoft Volume Licensing Program with Boeing. The Volume Licensing Program let corporate accounts pay a set fee per software license each year. This fee entitled them to any releases whenever they came out. It also let corporate accounts spread the budgets for software more evenly from year to year versus dealing with unpredictable release cycles. The Volume Licensing Program was groundbreaking within the personal computer software industry.

Another significant change that Microsoft innovated for personal computer software companies was made in selling to corporate accounts. The emphasis was placed on having direct relationships with them. Until this period, the majority of all software purchased by corporate accounts was sold through distributors and large resellers. Many personal computer software companies had relationships with distributors and resellers, yet not with corporate accounts. This put us all at a significant disadvantage to companies like IBM which not only talked directly to corporate accounts but often had an office on the premises of large corporate accounts. By creating a sales force that directly worked the largest corporate accounts, we could get information faster and accurately to and from this key target audience. We were able to directly influence the sale and not rely on indirect channels to make everything happen. We did this while continuing to sell through key strategic resellers who worked with our teams for our mutual benefit.

Lloyd Wilhelms (top right) and the PacWest Team in 1989

I was asked to do my own version of "worst to first" in the Western US region. The GM that I was replacing had a sexual harassment suit against him. And the region was behind plan, ranking last out of three regions. Rich estimated that I would be needed in Los Angeles for three to five years, after which the plan was to move me to Redmond, Washington, to Microsoft's head office.

Our Western region offices were first located in El Segundo, California, near many aerospace companies. We then moved north to the Howard Hughes Center just minutes from LAX airport and situated on the busy I-405 freeway. I should have rented out my office since I was on the road quite a bit. I had a circular corner of the building, and our floor had an outdoor terrace with a great view. It was prime LA real estate.

Microsoft was starting to make headway into corporate America. Microcomputers and the vision that Bill Gates had espoused were coming to pass. One day the vision became surreal.

Our office at the Howard Hughes Center was located in the Wang building. Wang Laboratories was the dominant force in specialized office equipment, with their dedicated word-processing machines. At its peak in the early 1980s, Wang had annual revenues of $3 billion. As we arrived to work one day in 1989, Wang employees were flooding out of the building with all of their office possessions in cardboard boxes.

"What happened?" I asked one of the box carriers.

"PCs are replacing our dedicated Wang machines" came the response. "We all just got laid off."

I didn't ask anymore questions. The displacement and upheaval our software and personal computers were already having on the world was physically demonstrated in front of me. The "safe" paths being chosen by these so-called established businesses had led them to the end of their roads. It became even clearer to me that day that businesses, in order to survive and keep moving forward, need to keep reinventing themselves.

- Do you have the right team on your most important accounts? Can you grow your largest accounts even faster by becoming strategic thinkers with them as opposed to being just another vendor?

Learning from Others

Success often comes by learning lessons from other industries. Laurie, Scott Oki's girlfriend and later wife, worked in the cosmetics industry at an executive level. The way in which the cosmetics industry staffed department stores with part-time people was modeled by our Area Sales Representative (ASR) program. Understanding how the cosmetic companies created a larger presence and served their retail customers and end users with their part-time staff was adopted as a strategic way for us to increase our presence and service level in the marketplace. We hired a part-time staff and taught them to be frontlines to all of our resellers. The rising stars in our ASR program went on to full-time work and even executive-level management in the industry.

Another key lesson came from a European consulting company that spoke at one of our sales meetings. These consultants emphasized measuring the actions that get you the behavior you want versus measuring "rearview mirror" statistics. Looking at traditional financial sales reports tells you how you have done. That's looking in the rearview mirror. However, it's much better to figure out the actions that drive you to the results you want. Measure and report on those statistics with your team.

This lets you get much closer to real-time results and incentives that create the behaviors that lead to success. The cosmetics industry measured its in-store representatives that way.

In order to maximize your success, you need to maximize the share of mind you get with your audience. Our plan: Do enough presentations and demonstrations to educate the resellers, and you will get their mindshare, and that will lead to more sales. We did exactly that with the ASRs. Our focus was on in-store visits and presentations to educate the retail software sales people. We measured activity statistics, including number of visits, presentations, and demonstrations. These statistics determined success. What we paid our part-time ASRs was based on these metrics.

- What can you learn from other industries that can help your business? Are you reinventing your business on a regular basis to stay relevant? Or are you letting someone else do it to you?

TEN

Train, Train, *TRAIN*

One of the principles of success that Microsoft instilled in me is the importance of training. I had learned this even before joining Microsoft, when I worked at Great West Life Assurance. GWL viewed training as key to the success of their business. Part of the training implemented there was computer-based. As microcomputers came into the company, I was tasked with a project to assemble a complete training program to help GWL employees learn how to use these new machines. Computer-based training (CBT) from a company called CDEX ran on Apple II computers. Leveraging CBT let us bring the project in ahead of schedule and under budget.

March 1, 1985 marked my first day as an official member of the Microsoft team. Rich Macintosh believed in training. On April 25 that year, he had me attend Time:Text from Priority Management Systems. In fact, Rich had everyone on the team take this training. It became our standard way of managing and communicating information. We used their project management templates, communication planners, meeting planners, travel planners, and more. Such tools gave us a common language and process for getting our work done. The lesson we learned was that having a standardized system creates leverage and speed. Leadership expert Zig Ziglar has said, "It doesn't matter which system you use, so long as you have a system." And having the entire team on the same system contributed to our success.

We were a young team "drinking from the fire hose," and we needed every advantage we could get. Learning the value of priorities—what should get done and how to prioritize tasks—became a life skill.

By the time I got to Los Angeles with Microsoft, I could no longer do everything that was on my to-do list. Breaking through that barrier was more difficult than I had imagined. I hated to see balls dropping. I pictured *The Lucy Show* episode where Lucy and Vivian are working in a candy factory. The candies are coming down the conveyor belt faster than they can handle them, hitting the floor and eventually creating a total mess and a room filled with chaos. That's how I felt at times. Nevertheless, the skills I learned from Priority Management helped me assess what should be done as well as to determine what could be prioritized lower and what could potentially never be done (so dropped from the list).

Priority Management was not the only training at Microsoft that we took. In fact, it was just the beginning. Microsoft continually invested in its employees. We took Acclivus Sales Training, Communispond Presentation Training, training on human potential and goal achievement from Jeff Goforth, project management training, and custom training modules developed for our executive sales leaders by Bill Meyer. We brought in Stephen Covey to train us on his *Seven Habits of Highly Effective People* methodology. And we hired Neil Rackham to teach us his major account selling approach, SPIN Selling.

Microsoft even invested in training its customers. We built Microsoft University, and we operated it in nine locations, including places outside of the United States. MSU was not an internal training effort. It was focused on our customers, primarily customers in corporations. Our largest customer was Aetna Life Insurance. We offered training on systems architecture, programming to train and convert the armies of COBOL developers that existed in corporate America to be able to use programming tools on microcomputers, and much more.

Doug Martin as the Church Lady of "Saturday Night Live"

Moreover, Microsoft made training fun. For example, *Saturday Night Live* and other timely social memes were often brought into our national sales meeting training events. One of my nicknames was "Jaws" (with a last name of Jaworski, that was a pretty easy one to get). I had the pleasure of dressing in a shark suit and replaying the famous "Candygram" line from SNL as I stepped to the microphone. (Unfortunately, I was unable to locate a photo of this event, so you will simply need to picture me in a full-body shark costume. It was awesome!)

We conducted many training sessions as a standard part of every national sales meeting. We used traditional training methods, as well as game scenarios, such as playing *Jeopardy* or modeling other popular game shows where the questions were all related to knowledge we wanted to impart to our sales team. This made the training fun and memorable. It ensured we developed consistent messaging across all of our team. It helped create excellence in the Microsoft organization.

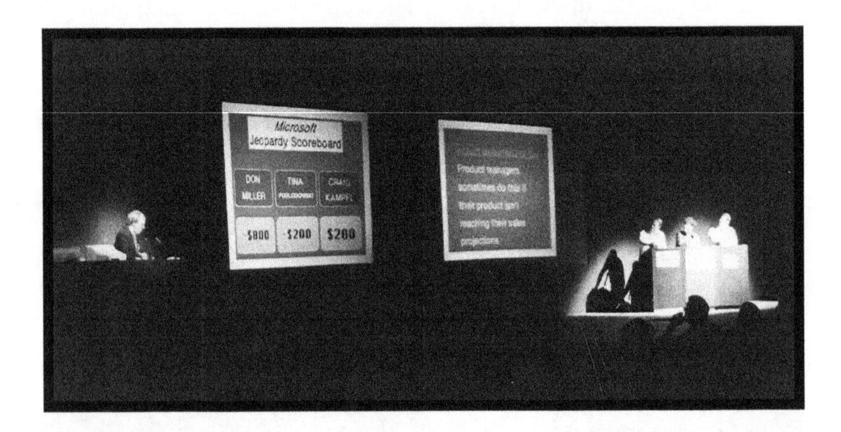

Training is usually the first thing that gets cut when budgets are tightened. It should, however, be the last thing that goes, if it gets dropped at all. An ongoing and consistent investment of training in your people is a difference-maker, especially in these times where so many companies run their teams and organizations leaner than ever before. Through training, you can get a young team to excel in a highly competitive market space. You can show your commitment to your people by investing in their personal growth and success.

How do you know what training is needed in your team? One quick assessment approach that I learned during my tenure at Microsoft was to determine if a gap in employee performance was related to will or skill. Will, meaning attitude, resulted in investigating motivational factors in the environment, in the person's personal and professional life, and so on. Skill, on the other hand, was addressed via training initiatives and skills based one-on-one mentoring.

- Are you invested in your team, offering them training opportunities to further their own skills and better your business overall? Do you perform regular assessments of your team to determine areas where training can make a difference? Do you have motivation gaps (will) or a gap in skills that training can address on your teams?

Choose Joy

Microsoft training included special regional events. On one such occasion our team was on the islands in the Puget Sound, west of Seattle. A great consultant who later became a full-time member of the Microsoft team, Doug McKenna, was leading our team in a workshop. In a matter of minutes, Doug changed from workshop leader to race-car driver, whisking me from beautiful Port Ludlow back home to the mainland.

My grandmother had not been well. I received an urgent message from Susan. I called her back immediately and heard her crying on the other end when she answered. I immediately thought my grandmother had died. The response from Susan was unexpected. She relayed that my dad had called, and he had not mentioned his mom. Instead his focus was on my brother, Wayne. Wayne received a call from the doctor's office following tests he had done a few days before. The tests were sparked by Wayne's persistent cough. The tests showed that he had leukemia (AML). The disease was so far advanced that the doctors didn't know if Wayne had more than a few hours left. I crumpled to the floor and started sobbing uncontrollably. My brothers and I are very close. Wayne had married Kathie, his beautiful new bride, just months before. I was in shock.

Doug raced me back home. While he played driver Mario Andretti, I called airlines and tried to coordinate schedules with my brother Gerry who lived in Roseburg, Oregon. Could we get back in time to see Wayne? Thankfully, God gave us time.

In the parking lot of the St. Boniface Hospital in our hometown of Winnipeg, Manitoba, Canada, I took a break and prepared myself for another visit with Wayne. He was dying. We were playing "Beat the Clock" in real life, working with the doctors to find a way to save his life. I was reading *Holy Sweat* by Tim Hansel. Tim had been a mountain climber. He suffered a serious fall and was destined to live with severe back pain for the remainder of his life. In spite of this, he said, "In every moment in life we can choose joy or we can choose misery. I choose joy." He wasn't talking about being happy. It was a deeper emotion. *How can I choose joy when my brother is dying?* I asked myself. In prayer, I determined I would serve my mom and dad, Wayne's wife Kathie, and the rest of us who were still in shock and mourning.

Rick, Gerry, and I—Wayne's brothers—passed the first-level test as potential bone marrow donors. That gave us a little hope. After Wayne underwent several rounds of chemotherapy, which caused his cancer to go into remission, we had even greater hope. But while his immune system was wiped out from the chemo, Wayne caught an airborne infection—the kind that you or I wouldn't even notice that our body had battled with and won. Yet, without his immune system functioning, the infection took hold and spread through his entire system.

Forty-six days after I had received the call about Wayne's illness, he was called home to be with God. The day was May 3, 1990. After all these years, I still miss him and think of him often. The lesson I learned during that time became an anchor tenet for the remainder of my tenure at Microsoft and beyond. In every moment of life, we can choose joy or misery. I choose joy.

- Choose joy.

ELEVEN

Bill Gates: *LIFELONG LEARNER*

One of the most asked questions I receive is, "What is it like to work with Bill Gates?"

Many stories have been written over the years about Bill's intensity. He is super smart, way ahead of the average person. He quickly gains an understanding of the topic at hand and fast forwards in his mind to conclusions and questions. It was common for presenters to start their presentations and find Bill already flipping ahead in the handouts, sometimes twenty or more pages ahead, and then questioning them on what he was reviewing.

Bill, with great emotion and angst, would say things like, "That's the stupidest thing I have ever heard!" Some presenters took comments like that personally, but Bill didn't mean them to be taken that way. The best way to respond to a strong statement from Bill was to go through the following thought process:

1. Either he is right.
2. Or I am right and have not convinced him or communicated effectively.
3. Or the real truth lies somewhere in the middle.

Then, as quickly as possible, ask Bill questions to get inside his head and determine which of the three scenarios was occurring. Finally, respond accordingly.

These interactions, which also usually happened during business reviews, proved to be incredible learning experiences.

Bill was a strategic thinker. He always had a five-year plan. As he approached the next milestone, he would goal set to the next plan period. Bill would take a week aside as a "Think Week." During that week he would read papers, explore software, and basically unplug from the day-to-day business of Microsoft to think long-term. This was not vacation for Bill. It was a time he took for deep thinking. It usually resulted in one or more strategic memos to the team and sometimes to the whole company.

As Microsoft expanded around the globe, Bill recognized the way our minds work and took advantage of that. He put a picture in his garage of the world map and said that would allow him to see and absorb it, bit by bit, every time he came to and left his home. He was always learning and expanding his own horizons. That motivated all of us to do the same. I believe that is why he shared stories like this with us.

Bill also surrounded himself with smart people. He let them challenge him. The best ideas then rose to the top. He did not have to be right. He simply wanted the best answers to emerge.

Bill Gates also understood the value in short-term strategic decisions on the road to bigger, longer-term strategic moves. For example, the "embrace and extend strategy" was used on many occasions to gain Microsoft a strategic place in the market. Consider. Microsoft adopted and licensed Lattice's C compiler as Microsoft C. Microsoft also adopted Microrim's RBase database platform and sold it as Microsoft RBase. As Microsoft learned the marketplace and the needs of customers, the company eventually developed its own solutions. On many other occasions, Microsoft recognized the industry leader, acquired the company and the development talent with it, and then furthered the platform internally. For instance, Forethought was acquired to bring their PowerPoint software to Microsoft. PowerPoint first appeared on the Macintosh, and four individuals developed it. The program was then ported to the PC platform. It continues to be a standard today.

Brian MacDonald's Project became Microsoft Project. Brian was then tapped to lead the creation of Microsoft's desktop information manager, Microsoft Outlook. People like Brian played significant roles multiple times in Microsoft's development history. Brian left Microsoft in 2001. In February 2007, he met with Microsoft's leader of BING search, Satya Nadella. Brian wrote a ten-page paper on opportunities and challenges for Microsoft's Search business. This led Nadella to rehire Brian to work for Microsoft.

Former US President Ronald Reagan had on a plaque on his desk that read, "There is no limit to how high you can go, as long as you don't care who gets the credit." He had a second plaque with an equally important message: "When the best leader's work is done, the people will say 'We did it ourselves.' " Frank Gaudette's statement rings true: "People are our greatest asset."

- Do you have a "Think Week" built into your annual calendar? If not, how soon can you work one in?

How to Gain Over 700 Hours a Year in Personal and Professional Productivity

Have you ever lived in your car? I felt I did, logging a minimum of three hours a day during my Los Angeles tenure at Microsoft. Driving interstate 405 to and from work every day, passing through the 405/101 interchange, was usually done at a snail's pace. It is said that more cars pass through that intersection point every day than at any other location in the US. Like the movie *Office Space,* a person in a walker could often pass through that point as fast as any car. "Sig Alerts" were too often the norm, meaning traffic was basically at a standstill. At first it was surprising to find yourself in a traffic jam as late as midnight in LA.

With that kind of time spent in my car, I decided to turn the experience into my personal learning lab. Earl Nightingale and Lloyd Conant were the leaders in personal and professional development at the time. Their company, NightingaleConant, offered a huge catalog of audio training and books on tape. These programs educated me daily. And with all that drive time, I was able to complete their entire catalog. I learned that you can take almost any obstacle and turn it into a benefit, if you want to.

Statistics on the radio news, such as "Children in Los Angeles have 15 percent less lung capacity than children in clean air cities," and talk-show humor discussing LA violence by joking that our popular bumper stickers were "Cover me while I change lanes" and "Honk if you are reloading" had us ready to move. We enjoyed the annual Disneyland passes, taking every visitor who came to town to the Magic Kingdom. Yet the final straw came one evening at 8:30. I had the kids with me, and we were stopped at a red light across the intersection from a major shopping mall. Traffic was busy. I glanced over to my right as people were surrounding a body lying on the pavement near an ATM machine. There was a young man in a pool of blood who had just been shot as he tried to make a withdrawal. This murder happened in broad daylight on a busy street corner. LA felt out of control for this Winnipeg boy.

One and a half years after we arrived in LA, Rich called and told me, "You've done everything we needed done. Now I need you at Corporate." I felt some relief, but his timing wasn't quite right. Susan and I were expecting our fourth child in September of 1989. She told me we were not moving while she was pregnant (again), and I passed that on to Rich. Jonathon arrived September 19. We moved at the end of that year. Susan and I were happy to get our family to the beautiful Pacific Northwest and away from LA.

Living in Bellevue, Washington shrunk my daily commute to Microsoft from three hours to eighteen minutes. I was elated. We soon built a home, and once it was ready, we moved into it. Our residence was now in Redmond, and this was where Microsoft's headquarters was located. Now my commute was down to just five minutes! Simple decisions about where you live can have a radical change on your life. Not only was my commute shorter but increased family time became a regular reality and lunch with Susan and our children became an option we enjoyed many times.

My new role kept the learning fire hose going for me. I was now the general manager of US Sales Operations. Eleven direct reports, each with a different set of job responsibilities, emphasized the critical importance of my role. Our team led customer service for sales, inside sales, the systems engineer program, SE training, field sales training, the operations review committee (which coordinated all launch efforts between product marketing and the field sales organization), sales and systems reporting, and even more. There were over five hundred people in the US Sales Operations organization at the time.

Scott Oki and Rich MacIntosh taught us a great deal about strategy. We did SWOT analysis on all competition, analyzing strengths, weakness, opportunities, and threats. We brought in leading thinkers, such as Neil Rackham and his SPIN sales training, for major account sales. Stephen Covey had released his book *The Seven Habits of Highly Effective People,* so we brought him in to address the troops. We reverse engineered the profit-and-loss statements of private and public companies. For private companies this meant gathering clues from every possible source or public statement made. Then, we worked to find their Achilles' heel, focusing on points that emphasized our strengths while highlighting their weaknesses. The bigger this wedge and the more customers valued the points we made, the faster we could beat our competition.

We used this approach to go after companies dominant in our industry, including Lotus who owned spreadsheets with "1-2-3," Novell and their networking products, and Borland.

- What learning opportunities can you build into your environment?
- How do you leverage the strategic wedge and create advantage for your company? Do you conduct SWOT analysis on your company and your competitors?

TWELVE

Know Who *YOU ARE*

Microsoft is an API company at its heart. API stands for application program interface. Computer programmers can write to a program interface and gain access to the capabilities of the underlying system. Consumers benefit from APIs, even if they don't know what they are. Microsoft's APIs were its "crown jewels."

Borland, one of Microsoft's chief competitors, had set their eyes on Microsoft's crown jewels. This led us at Microsoft to view Borland as a major threat, and we went after Borland accordingly.

Protecting the Crown Jewels

In order to understand just how serious Borland's move for the crown jewels was, it is important to understand some technical background. Stick with me through these next few technical pages as the analogies that emerge may prove very helpful to your business and industry.

Strategic Information

The operating systems—DOS, OS/2, and Windows—let programs talk to their program interfaces and gain access to a wealth of capability. The change to graphical user interfaces like Windows and the Mac had a huge benefit beyond the visual changes. Programmers were able to use common libraries to display menus, windows, and dialog boxes in consistent ways across programs, even programs from different companies. Under the hood, rather than each programmer for each program having to write printer drivers, screen and font drivers, external peripheral drivers, and the like, the operating system developers do that once and provide interfaces that are easily called by all programs. Early software developers had to write hundreds of printer drivers alone. Leverage makes software affordable, more reliable, faster to market, and flexible. When a new printer comes to the market, only the operating system needs to understand it in order for programs to instantly recognize it and take advantage of it. Everyone wins.

When I first joined Microsoft, it had one—and only one—universally successful program. That was DOS. DOS stands for disk operating system. DOS was based first on a floppy disk and then later on the hard disk of the computer. DOS was a 50K program. Fifty thousand characters brought the computer to life and gave it the ability to handle instructions from software programs and interface to hardware.

Even though DOS provided the engine for much of Microsoft's early growth, Microsoft didn't create DOS from scratch. IBM came hunting for an operating system they were going to use for the new personal computer they were bringing to market. Gary Kildall's CP/M (Control Program for Microprocessors) operating system was the original operating system that was going to power the IBM PC. John Akers was head of IBM at the time. He was on the national Board of Directors for the United Way with Mary Gates, Bill's mother. During the process of evaluating IBM's options, Mary recommended that John talk with her son's company. Historians tell us that Gary Kildall made a mistake. Rather than meet with IBM when representatives came to town, Gary went flying (he was a pilot). His absence opened the door to an alternative solution.

Tim Paterson made a product called QDOS. Microsoft bought the rights to his disk operating system for $50,000 and then modified it to suit their needs. This became MS-DOS (Microsoft Disk Operating System), and it was licensed to IBM as PC-DOS. Tim left Microsoft and started his own company, Falcon Technology (aka Falcon Systems). He was contracted by Microsoft to write a port of MS-DOS to new MSX computers that Microsoft was building in partnership with ASCII for Japan and other markets using the ASCII Microsoft 8-bit computers. Microsoft acquired Falcon Technology outright in 1986 to have royalty-free licenses to all versions of MS-DOS. DOS then became Microsoft's bread-and-butter product.

After several versions of DOS, Microsoft believed a new version called OS/2, a joint development effort with IBM, would dramatically expand the capabilities for the new hardware that was coming into the marketplace. Yet there were problems that even Microsoft could not overcome. Like a bad Dilbert cartoon, IBM's engineers were paid and given bonuses for the number of lines of code they wrote. In software development, less code, not more, to accomplish a task is usually better. IBM also had to support a huge legacy of devices in the marketplace. All of this slowed down the development of OS/2.

At the same time, Microsoft was developing a new operating system called Windows. Both Bill Gates and Steve Jobs had seen the work that Xerox was doing at its Palo Alto Research Center, Xerox PARC. Both leaders were inspired by the windowing interfaces being developed, input devices such as mice and more. Microsoft announced Windows in 1983, ahead of the release of Apple's Macintosh in 1984. But Windows took longer to bring to market than Microsoft had anticipated. It didn't release until November 1985.

Even with the release of Windows, Microsoft and Bill Gates kept the faith for OS/2. In 1988, Microsoft released OS/2 v1.1, which offered true multitasking, enabling the computer to do multiple tasks at the same time. On other personal computer operating systems, when an application went to the background, it effectively froze until it came back to the foreground. For this reason, inside of Microsoft, most people, including Bill Gates, first believed that OS/2 would be the major personal operating system used by companies. Yet the burden of device support and the two-company development issues resulted in Windows having flexibility and speed that OS/2 would never enjoy. Eventually the true multitasking capabilities were added in Microsoft Windows.

At the time of OS/2's release, Steve Jobs was in his hiatus period from Apple. Released from the company he had founded, Jobs started NEXT. He built the NEXT operating system on an operating system called Unix Mach. Even Bill Gates said Steve chose well. Mach was, in Bill's opinion, one of the best versions of Unix available at the time. It too offered true multitasking, much like OS/2. Bill predicted that Unix would have a place on servers and actually spoke publicly about this. He also said he did not believe the market had room for another desktop operating system beyond what Microsoft and Apple were offering, and he predicted that a Unix desktop would not become a third desktop, even if it came from IBM or NEXT.

Time has proven Bill correct on both his Unix predictions. Linux, a Unix derivative written by Linus Torvalds, is a dominant server operating system for servers running the Internet today. Apache and other modules that run on Linux and Unix power the majority of web servers. The market did not make room for a third standard PC desktop. Steve's NEXT was unable to crack the market open for itself, so Steve found another path.

When Steve came back to Apple, the NEXT operating system was acquired and became the new Macintosh operating system. So while the NEXT O/S failed to become a third desktop standard, it replaced the former Macintosh desktop O/S with a more robust and full multitasking O/S. Windows flexibility and developer community support gave it speed and flexibility that led it to become the other dominant standard over OS/2. The multitasking intelligence and networking capabilities of OS/2 were migrated to Windows starting with Windows 3.11, also known as Windows for Workgroups, and then fully with Windows NT (New Technology).

The initial development of Windows NT was led by one of Bill Gates' prized recruits, Dave Cutler. Dave was recruited from DEC where he was the operating system genius. Bill said Dave was one of the smartest people he knew. Bill recognized the importance of having the best, smartest, and deepest thinkers on his team. This created what he called a "positive spiral": great people attract other great people and soon you have this positive spiral going on.

Personal Software Corporation, in 1982 renaming itself VisiCorp, is a great demonstration of a "negative spiral." Personal Software Corporation distributed the first spreadsheet for personal computers, VisiCalc, which was written by Dan Bricklin and Robert "Bob" Frankston. Dan and Bob's company was Software Arts. They wrote VisiCalc and used Personal Software as the distribution company for it. It sold seven hundred thousand copies for the Apple II. It was the first program that turned personal computers from hobbyist machines to professional tools for business people. Personal Software Corporation also developed VisiTrend, VisiDex, and VisiPlot. Mitch Kapor, a talented young engineer, was behind several of the VisiCorp products.

VisiCorp also had the first commercially shipping PC windowing system. Virtually no computer users today are even aware of its existence. It was called Visi On. Citation Software, where I worked prior to joining Microsoft, had the Canadian exclusive for VisiCorp products. The Visi On Applications Manager was sold in the US for $495. Then a consumer would need to pay an additional $250 for the mouse as well as purchase applications, such as VisiWord, for word processing. Consumers could not do anything without Visi On and the optical mouse. This setup led to VisiCorp pricing itself out of the market, and this opened the door to Apple and Microsoft.

First to market doesn't always win. First impressions can help or hurt you. In fact, even changing or lowering price once the market has an understanding and positioning of your product does not necessarily change the public's perception of the product.

The other interesting lesson from VisiCorp is about arrogance and leadership style. Mitch Kapor and other leading developers disliked the autocratic style of VisiCorp's President Terry Opdendyk, so they left. Mitch wanted to build a new version of the spreadsheet. Terry wasn't interested. Even though others on the executive team suggested they block Mitch from creating a new spreadsheet software program when he left, Terry didn't feel it was worth the effort. He even derided Mitch's skills. That arrogance led to Terry executing a one-page agreement that allowed Mitch to create a new spreadsheet without any ownership or requirements to Personal Software. Mitch formed Lotus Development and wrote 1-2-3. Mitch's business acumen almost single-handedly brought down VisiCorp. In fact, once 1-2-3 started to gain success, VisiCorp filed suit. The one-page agreement, signed by Terry, let Lotus 1-2-3 stay independent. The saga had more turns. VisiCalc, VisiCorp's flagship product, eventually ended up owned by Lotus as VisiCorp sued the original creators of the groundbreaking spreadsheet program and their company, Software Arts. A countersuit followed. VisiCorp lost the battle and, eventually, the war.

Arrogance comes before a fall. This theme has been repeated historically many times. And Microsoft would be no different.

The Fight: Microsoft vs. Borland

Now, back to our API story and Microsoft's crown jewels.

Borland was enjoying some success with its application products, including Paradox, its spreadsheet program, and ANSA, which was its database program. Borland also started making inroads with programming languages. Borland started talking to developers about using its new super APIs. These so-called super APIs would let you write your code to Borland's APIs. In turn, Borland's APIs would talk to Windows APIs and, in theory, could address other underlying API sets.

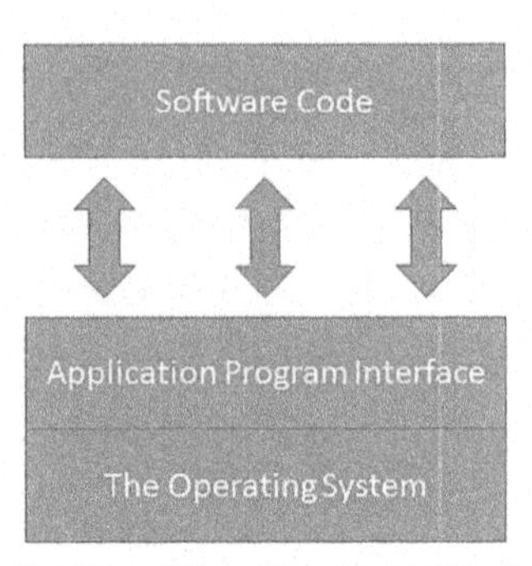

Traditional Operating System Stack

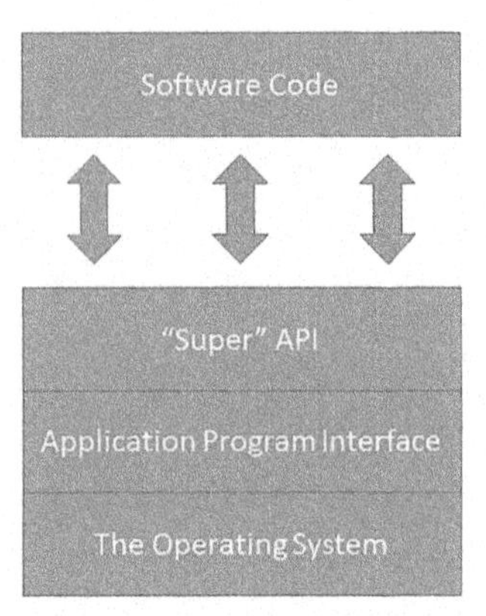

Borland's Attempt to Insert a Layer

With this move, Borland had effectively declared nuclear war. If they were successful getting developers to switch to their super APIs, they could easily move their programs to any other operating system—their own or Unix or anywhere else. They could move their intermediate layer to sit on another operating system and take the base of applications and all the associated benefits with them, effectively devaluing the Microsoft O/S. They were going for Microsoft's crown jewels.

In my early years at Microsoft, WordPerfect was number one in word processing. WordPerfect, Wordstar, Multimate, IBM DisplayWrite, and XyWrite were all purchased by more consumers than Microsoft Word at the time, and Volkswriter was close as well. Through a series of strategic marketing efforts, Microsoft Word climbed the ladder from number six to number one. Key marketing programs included a Word Sampler disk to give people a hands-on experience with Word and the Microsoft Challenge. The Challenge focused primarily on WordPerfect, often not even acknowledging the others as Word ascended to the top position. The Microsoft Challenge was like the Pepsi Challenge. Pepsi's campaign was a taste test to have consumers compare Pepsi against Coca-Cola.

While character-based Microsoft Word had a long, slow, uphill climb, the release of Windows offered a window (pun intended) where consumers would be experiencing a new interface to their Word processor. We studied the features that all word-processing users used and focused on being the best in the features that mattered most. Microsoft leveraged the strength of Word's position as a member of Microsoft Office and the common interface enjoyed across all programs.

This same formula was used to unseat the competition in spreadsheets. Excel was a full member of Microsoft Office on Windows. Combining our strengths with Office and common menus, the movement of all products to the new Windows environment, and studying the Lotus 1-2-3 user community to learn what they loved and what they wanted, we became as knowledgeable (or better) about the 1-2-3 user than the Lotus team was. As shared earlier, we did SWOT analysis to determine how to best position Excel. And the development team gave the sales team its biggest opportunity with a little magic.

Most spreadsheets calculated the entire spreadsheet before returning control to the user. When you had a very large spreadsheet, this could mean a significant wait before you could enter the next number or even move around the document again. Jabe Blumenthal led the Excel developer team and saw an opportunity to take a new approach: just calculate the area that was on the screen, understanding any of the dependencies for the cells on screen, and then give back control to the user while finishing the recalculating of the spreadsheet in the background. When this one feature came out, Excel looked like lightning and made Lotus look like a snail. The recalculation happened so fast that people would often change another number and put their finger on the screen to a cell that they knew would have to change when they pressed enter. They would literally laugh and marvel out loud when it happened instantly. It happened so fast, they usually tested it a couple of times to address their initial disbelief. Attention to these seemingly small details helped open-minded users see how Excel was going to give them back minutes and even hours a day in their use of this technology over any other spreadsheet. Jabe and his team also ensured that all the muscle-memory keystrokes of Lotus 1-2-3 could be used in Excel to get the same results. So there was no need to relearn the new program. Users could migrate to the mouse and other Excel features as they liked.

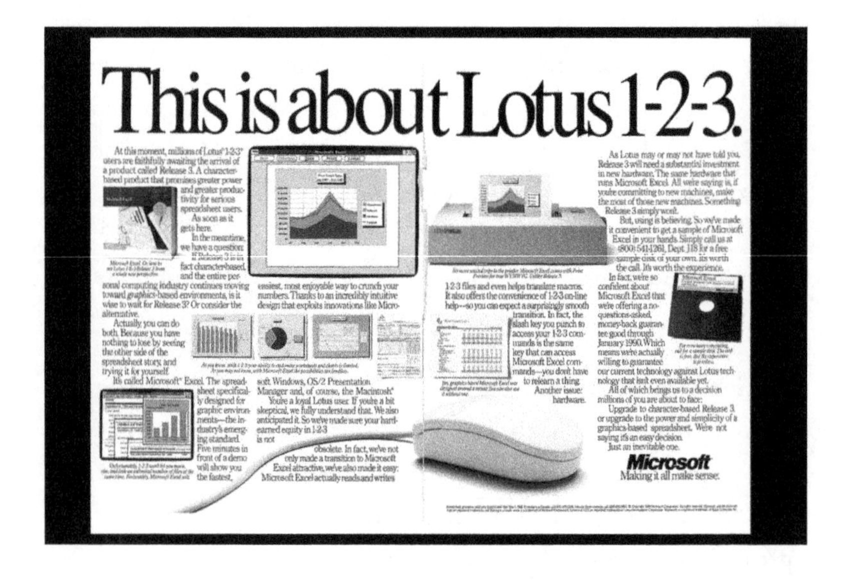

On the language front, Borland's promise of giving developers "super powers" by letting a line or two of code access a "super API" was attractive. It meant developers could get more done in less time. And it meant less debugging time. Borland was onto something that the developer community would find incredibly desirable.

Microsoft did a few strategic things to combat Borland's move. We looked at Borland's overall business and figured out what drove their cash success. We attacked on that front. Then Borland started a pricing war in applications. I don't believe they thought we would respond, but we did. And we knew our response would have a significant impact on Borland's financial capabilities. So we went hard at Borland. Be careful who you pick a fight with and what assumptions you make. We also created our own new APIs that gave developers additional capabilities. We kept the target moving forward so fast that Borland could not catch up. Being on Borland's language platform could mean you were behind if Borland didn't take advantage of all the new capabilities we were bringing out. We struck another blow when the super-talented Brad Silverberg left Borland and we snapped him up to join Microsoft. Brad became a leader in the company and later delivered the solid Windows XP platform for us. Numerous blows like these resulted in Microsoft keeping control of the API crown jewels and Borland eventually fading into obscurity.

A few key people make a difference. People write your company's software and run your business. Michael Phillips, a relative and dear family friend, is a retired business executive. His advice: You are only as good as your weakest employee. He elaborated, saying that the front-lines to your customers set the tone for who the company is in their mind. They define the personality of the company. One weak link can destroy a great company. Many companies overlook this and knowingly hire a B-player rather than going for a team of A-players. Microsoft started to justify this kind of action in the mid 1990s, saying it needed people as quickly as possible to meet all the company's needs as it continued to grow. Over time, however, this tactic creates chips in the culture rock of your company. It can eventually take a company down. One chip leads to another. A B-employee hires a C-worker, and later that C-employee hires a D, and farther down the trend goes. In the meantime, B-employees deliver B-results, and C-employees deliver at their level of expertise and experience as well. If you want an A-class company, you need A-class employees. Shortcuts will not deliver the goods, and they may derail your business.

In the world of software development, an A+ developer is ten times as effective as an A developer. A+ developers write better code, write it faster, and with fewer errors. Similarly, an A developer is ten times better than a B one. That means that a B developer is a hundred times less effective than an A+ one. C developers are not just a little worse than a good developer; they are a thousand times worse. Make too many compromises and you soon fall in the sad statistic of failed software development products. Rather than going in this destructive direction, it's far better to protect the culture of your company. As Peter Drucker said, "Culture eats strategy for lunch."

As a new filter, Microsoft started to require MBA degrees for many positions. I was at lunch years later with Ron Davis and Jim Minervino. Ron had run corporate marketing for Microsoft during the hypergrowth years. His thinking was key in many of the decisions that led the company forward. Jim Minervino led Microsoft marketing research. None of the three of us had an MBA degree. Dhiren Fonseca, our close friend who had also been a Chairman's Award winner and went on to be one of the top executives at Expedia didn't finish his undergraduate degree. None of us could have made it through the new hiring criteria. Yet all had played key roles in Microsoft's rise from worst to first. In fact, by that standard, Bill Gates could not have passed the company's criteria as he too had not completed his college education. The company's new filtering criteria for new hires did not guarantee the acquisition of better employees. At times it even backfired.

The war and rise to the top of the computer industry went on for several years. It was marked by numerous battles on all kinds of fronts.

The collective effect of multiple marketing campaigns, strategic marketing strikes, leveraging the position of our products as a suite, helping one success beget another, and hiring the best people as we grew, which led to the attraction of more great people, all worked together to help Microsoft achieve its objective and, ultimately, its mission of a computer on every desk and in every home running Microsoft software.

- What are your company's crown jewels? How do you protect them?
- Is your criteria set correctly to get you the best people?
- Are there any areas where you have chosen to compromise in your personal or professional standards? What can be done to protect your personal and professional culture?

Undercover: Project Copernicus

Probably one of the most interesting projects that I did at Microsoft was Project Copernicus. The goal of Project Copernicus was first and foremost to understand ourselves through the eyes of a customer. The secondary goal was to understand our competition through those same eyes. This involved benchmarking us against our competition. For the work to achieve its goal, I went undercover for three months. Three of my covers had physical offices:

- I was an IT manager for Century 21 Real Estate in southern California.
- I was also an IT manager for Great West Life Assurance Company in Denver, Colorado (US HQ). I even scheduled onsite meetings at their campus.
- I was an IT manager for Ogilvy & Mather in New York City—one of the largest advertising firms in the world at the time. I scheduled onsite meetings at their offices too.

All of these companies cooperated with our research. In exchange, we shared the results with them.

Going undercover could have given us access to extremely confidential information from our competition. We did not cross that ethical line. We took no beta or unreleased software from any competitors, even though it was offered to us. When we were offered pre-release versions of their software, we simply delayed with, "I am too busy right now. Thank you."

Philippe Goetschel, who was new to Microsoft, was brought on the project with me. Since I had been with Microsoft for many years and was known to many people within the company, Philippe would handle the in-person meetings with any Microsoft people. I handled phone conversations and also met with competitors. We also shopped retail, dialed technical support, dialed customer service, and called the main switchboards of Microsoft and all major competitors. We even measured response times for setting up meetings. Our goal was to understand who was servicing customers the best, and how we fared in comparison when we were not the best.

The results that came from Project Copernicus reached across the entire company. Recommendations included packaging, how we worked with customers in field and corporate support, and even how we answered the Microsoft corporate switchboard.

Until that time, there was no standard script. The most common response was, "This is Microsoft." Not very friendly. Not even personal. Microsoft was perceived by some customers as arrogant. This had to be fixed.

My dad was an entrepreneur. He taught my three younger brothers and me how to answer the phone. Any phone call to our home could be a customer for him. I took what I learned from my dad and applied it to Microsoft. So the new greeting format became, "Thank you for calling Microsoft. My name is Dave. How can I help you?" All callers were greeted and thanked. You knew you were talking to Dave. Microsoft had a face. And you were asked how you could be served. You were not a burden on our day. In fact, you were the reason we exist. That new greeting became the standard protocol for Microsoft.

When we presented the results of Project Copernicus to all departments across the company, an interesting thing happened. People agreed with all of the findings except the ones that related to their area. They felt we had nailed it in regard to the various areas of the company, except as it related to them making changes. Fortunately, we had the support of the executive management, and many of the recommended changes became the new way of business or at least influenced how we did business at Microsoft.

TAKE ACTION!

- How well do you understand the experience your customers have with your company? How well do you compare with your competitors? What changes can improve your opportunities with existing and potential customers?

THIRTEEN

Rewards Can Reward or *DEMOTIVATE*

In a one-on-one meeting with Rich MacIntosh, Rich commented that my performance continued to be very strong. I had received fantastic performance evaluations—all fives in Microsoft's performance evaluation rating scale. A new opportunity had opened up within the company. A new vice president was to be named to run Product Support Services, the group that provides all the technical support for Microsoft customers. The vice president position came with a raise and the eligibility for twice as many stock options at grant time.

Rich told me, "We wanted to let you know that you were our first choice. But we do not have enough female executives. We need a woman in the position. So Patty [Stonesifer] is getting the job. We want to let you know how much we appreciate you and all you are doing.

Your work is exceptional." I was also told that if I publicly repeated this, it would be denied.

Equal Employment Opportunity Commission guidelines are implemented by companies to try and reduce discrimination. Many companies set goals for increasing opportunities for women and minorities. In doing so, these goals can create new kinds of discrimination. As a result, the standard of merit often does not determine opportunity. Many types of discrimination in the workplace continue to this day. Perhaps you have experienced one type of discrimination or another. It never feels good. And while you technically have rights you can assert through EEOC processes, the invariable outcome, in my opinion, is a lose-lose situation.

I shared the news with Susan that evening. I felt like the Michelin Man after meeting with a large needle—totally deflated. While I appreciated the compliments on my work, I had been told that merit had not determined the recipient of the benefit. It was Patty who would step up to the VP level, receive a raise, and be eligible for twice as many stock options. Words of praise were nice. Achieving the new position and the increased compensation would have been far better.

I am forever grateful for the many blessings I received at Microsoft, and I continue to see Rich as the best person I have worked for in my career. Yet I would have preferred not having had that discussion at all given the predetermined outcome.

- If you have someone who is doing a great job, yet for reasons of "equal opportunity" (or any other) you determine to promote someone else, that is best kept private. It is not motivational, and it violates EEOC federal law.

- Some of you may disagree with me on this, but I'm going to say it anyway. Promote based on merit. Reward your top performers. Everyone should have the opportunity to advance based on their true merits.

Understand the Real Threats

On November 14, 1989, an unexpected invitation arrived from Jon Shirley, Microsoft's president. The "first Microsoft Management conference" was going to be held. The invite read:

> The purpose of this meeting is to tap a broad, experienced management group to discuss and propose solutions for the strategic and organizational challenges Microsoft will face in the coming years. A handful of people from each of the divisions of the company are being invited. We expect a total of about 35 people to participate including about six executive staff members. Executive staff members will serve as facilita-

> tors and information resources, rather than as participants in the discussions. Our aim is to hear from you and other key managers from all parts of the company. To kick things off, Bill Gates will give a talk on markets and technology; next, Frank Gaudette will talk about the financial model and performance of Microsoft. The meeting will then break into study groups of about five people each, to discuss and make recommendations on issues of importance to Microsoft.

In my study group, our assignment was this: define the biggest threat to Microsoft. Our answers included WordPerfect, Lotus and their 1-2-3 product, and, my answer, arrogance. The group laughed at my suggestion, but I was serious. Many of the great software companies had fallen when they became arrogant. Examples were all around our industry and included PFS, Ashton Tate, VisiCorp, and others. I already saw the early seeds of arrogance planted in various parts of Microsoft.

This was not the first time my suggestions created laughter yet proved true. In one strategic planning meeting, I suggested that the company should consider paternity leave and even a sabbatical for employees who had long tenure with the company. With the effects of a hard work ethic over many years, I suggested that we could see our most tenured people "dropping like flies" and leaving. Rather than taking my idea seriously, my phrase "dropping like flies" was repeated almost every session for the remainder of that weekend planning session to hysterical laughter and personal mockery. It was many years later that the company finally added paternity leave and sabbaticals, but only after many of the most talented and tenured people had long since left.

- Does your company have a paternity policy? A sabbatical policy? Should it?
- Have you identified your biggest threat?
- If you see arrogance building in your organization, snuff it out as quickly as possible.

FOURTEEN

Our Greatest *ASSETS*

Mike Hallman, a former executive with Boeing Computer Services and International Business Machines Corporation, became Microsoft's president after Jon Shirley resigned. He served from 1990 to 1992.

In late 1991, I received a call from Mike Hallman's office. Mike wanted to meet with me. In that meeting, Mike shared that the executive team was making long-term plans. They were looking at current leadership and assessing who could lead Microsoft in the future. Mike said that I was one of five people they seriously considered as a candidate to become president of Microsoft one day. In order to groom me for that possibility, they wanted me to run a full profit-and-loss area of the company. Normally that would mean sending me abroad to run an international subsidiary. As I had already lived abroad, working in international at Microsoft Canada, they felt moving me back to an international subsidiary was unnecessary. Instead they would put me into a P&L operation right in Redmond, Washington. Microsoft University was a full profit-and-loss operation. The lady who had been running it was leaving the company. The executive team wanted me to run Microsoft University to get that valuable P&L experience as part of my training. I accepted.

Microsoft University

By this time seven years had passed since I joined Microsoft. Many people knew me, but others did not. I wanted to let this latter group understand some things about me. I was introduced to the entire Microsoft University team in a meeting to kick off our relationship.

"My name is Dave Jaworski. I've been with Microsoft for seven years now. Many of you have heard a little about me. I'd love to know what you have heard, and I'd love to get to know you."

"We hear that you work really long hours and send emails late into the evening," commented one person.

"Yes," I affirmed. "And, at the same time, what you may not have heard is that I make every possible effort to have dinner with my family every evening. Then, as the kids head to bed, I often get back online and do email."

Then I added, "I am a Christian. I love Jesus. I'd love to know a little about you and your personal lives because I believe our lives are not segmented. Who you are at home impacts who you are at work and vice versa."

Many people have assumptions about what being a Christian means. Some of those assumptions are correct, while some are distorted stereotypes. Others are simply incorrect. I later thought of the Gary Larson cartoon where two deer are standing up in the forest talking to each other. One looks at concentric circles on the belly of the other and says "bummer of a birthmark." I would soon learn that I had painted a target on myself for those who had their own opinion of what being a Christian meant.

Tina Podlodowski had been the head of Microsoft University. In my briefings for taking the job, I had understood that the goal for MSU was to be profitable. Specifically, I had understood that the goal was to be plus 10 to 15 percent on the bottom line. I met with Tina and asked where the group stood. She told me they were "seven points behind plan." Seven percentage points less than 10 or 15 would still have the group on the positive side, just not as high as the stipulated goal.

I responded, "So you're in the positive. You're just not as profitable as you'd like to be."

"No. We're more like break even," she said.

This math was not working. I placed a call to the Microsoft Finance team and asked for a clear set of financials as to where the business stood. Finance came in and gave me that clear assessment, which provided the stake in the ground for where we were. I needed to ensure everyone understood our reality and our goals. By the time I received that report, Tina had moved on. The report from Finance was stunning.

MSU was *"minus* 70 percent" on the bottom line. This business unit did not need tweaking. It needed major surgery. The MSU business unit had lost over $5.58 million that year to date.

Just nine months later, our team at MSU had turned the tables and generated a significant profit. This was a mini worst-to-first story, yet it almost didn't happen.

Tina left me a stack of financial reports over a foot high on her desk. As I went through the many spreadsheets and accounting reports, I came across a "Dear Jane" letter from Tina's lover. It disclosed their lesbian relationship. I learned that Tina was leaving Microsoft to champion the gay and lesbian movement in the Seattle City Council. MSU had a large number of gay and lesbian people, and they had felt comfortable with Tina as their leader. Because I had identified myself as a Christian, many of them felt that I would be a threat to them and their career, that Christians judged them and would be biased against them. Many chose to judge me and assume what my beliefs meant.

What they did not realize is that I believe as Christians we are called to love first. In the words of a great Steven Curtis Chapman song, "God is God": "God is God and I am not." Jesus said the job of his followers is to (1) love God with all their heart, mind, and soul and (2) to love their neighbor as they love themselves (Matthew 22:37–39). Jesus did not tell us to pass judgment on others. Judgment is God's business, not ours. There are few occasions when judging others is justified in the Bible, and we Christians often get this wrong. Jesus says, "This is my commandment, that you love one another as I have loved you" (John 15:12). What people first need to know is that Christians care about them. For me as well, I like to understand a person's personal beliefs and their view of God. That, I have found, is a better place to start a discussion with them.

I was not at MSU to pass judgment on people's personal lives. In addition to doing my best to live out the calling to love, my work focus at MSU was most concerned about the task at hand—namely, our performance as a business unit. We were the poorest performing business unit in the entire US Sales and Marketing Division. And that had to change quickly.

Bill Gates, Steve Ballmer, and Frank Gaudette called me into a meeting to review the situation at MSU. They asked me to write a two-year business plan to turn around the MSU business unit. And then Steve added emphatically, "And don't let it take two years!"

Working with our leadership team at MSU, we reviewed MSU's performance in all of its nine operating centers from top to bottom. We looked at what was working and what wasn't. We assessed each team member based on the merit of his or her contribution and value to the organization.

MSU had previously won awards for packaging. That was no longer going to be success criteria. Our customers wanted the training in advance of the release of our products. They would rather receive it sooner in a brown paper bag, poorly photocopied, and wrapped with bailing twine. That was better than waiting for training that would come out months after a product was released, even if it looked beautiful.

Worse still, the MSU team had been told the business unit was performing well. The internal awards given to the team by MSU management served as an affirmation to that. The actual financials were not shared with the team. So the facts that I brought from Finance and my review with Bill, Steve, and Frank varied so much from what the team had believed about themselves that they viewed me as misinterpreting the information. They debated with me as they grappled with such a divergent picture of their unit's alleged success. It took some time before they understood the true situation at MSU.

In order to accelerate our speed to market and to meet the customer requirements of having the training in advance of product releases, we placed our training developers into the product groups themselves. They literally spent time officed in the same buildings as the people who were writing the code. They became part of the meetings that product development teams held. By becoming insiders to the process, we could meet the needs of our customers.

Learning from the success of Novell and their certified training program, we had Bill Lane, a former Novell employee, lead the creation of what became known as the Microsoft Certified Professional program. MSCP became a standard that businesses used to assess the skills of their partners and staff. We also put together a training program to get everyone on the same page. I wanted everybody to understand their potential and that of the organization. We worked with the Pacific Institute and its CEO, Lou Tice, to get everybody on that same page and moving in the same direction. The goals of the program included helping people understand comfort zones, goal setting, leadership thinking, creativity, and much more. It was a comprehensive program. Nevertheless, I soon learned that the MSU team viewed it as highly offensive.

One day I had a knock at my office door. It was a woman from Microsoft HR.

"You're going to get sued, and you're going to get us sued," she said, quite upset.

"Why?"

"The training you are having everyone go through is resulting in complaints to HR. The words you are using in the training are highly offensive. There are two words that are particularly offensive to your people."

"And what are those words?" I asked.

She told me that the first word was God.

"Okay," I replied. "I can understand how some might find that offensive. What is the other word?"

"Family," she answered.

"Really?" I was stunned. "Has Microsoft become such a dark place that family is viewed as a bad word?" In that moment, I prayed, *Lord, give me an answer.* He did. No sooner did I think the question than an answer appeared in my mind. It was as clear as day. I knew what the problem was. "No problem," I said. "I'll take care of it."

She asked what I was going to do. I told her that I would make changes in the program, and she could let me know if the threat of a lawsuit continued.

What God revealed to me in that moment was that it was not the words but rather the delivery that was offending people. We immediately canceled the use of Lou Tice's Pacific Institute program. Lou Tice was off the charts in the "driver" personality dimension. He had an energetic and intense personality. If you too were an off-the-charts driver or a highly verbal person, you were usually fine with that. However, if your personal style was more reflective and you preferred written information and taking time to digest it, you felt steamrolled by an overly assertive person like Lou.

I had worked with Bill Meyer, who had actually co-created the original Pacific Institute curriculum with Lou many years earlier. Bill had trained our US sales force with a program we called "Leading Microsoft in the 90s." (This sounds so old now, but it didn't then.) Bill was a National Speakers Association presenter and had spoken to over one hundred thousand people at the time. His style was flexible. The way in which he presented was accepted by people of all styles on the DISC spectrum, which delineates four main personality types—dominance, influence, steadiness, and compliance. Bill and I defined a curriculum from the collection of materials that he regularly spoke on. We called our program "Unlocking Our Potential." Every unit of this program was backed by texts found in the Bible. I wanted a program with integrity and values as its base.

The program with Bill as its presenter worked well. Not only did we train people, giving them all they needed, but we never got sued—not MSU or me. The majority of the MSU team came and told me what a great program it was. "Unlocking Our Potential" gave us a common vocabulary. It helped us quickly get on the same page. Once aligned, we were unstoppable!

Now fast-forward through the rest of that year. Microsoft University was the only group that made their business plan every single month of the fiscal year. And in less than a year, we achieved profitability. We had months where we were 34 percent over plan, and by October our year-to-date results showed a profit contribution average of 20.2 percent. In the US alone, we achieved a $5.6 million swing from a $4.3 million loss to a $1.3 million profit.

While at MSU, I wanted to respectfully leave the door open to share my faith whenever anyone was interested in learning more about it. I also understood that in the corporate environment, preaching was unacceptable. So I decided to keep doing what I had chosen to do years earlier. I decided I would put my Bible on my office credenza in full display, an action I have continued throughout my career. Anyone who asked me about my Bible or made any comment about it had effectively given me permission to talk about my faith. If no comment was made, I was hopeful that the example I set in how I lived my life would speak for me. I set high goals for myself, and I strove to attain them while not leading people off course when I messed up. You see, I believe we all, including me, need a savior. I am thankful for God's grace and forgiveness through Jesus Christ. (Time for a plug for one of my favorite songs of all time: "Forgiveness (Heart of the Matter)" by Don Henley of The Eagles. Check out the lyrics. Beautiful.)

Over the years, many people made comments about that Bible sitting on the credenza. Some started cynically, "Is that there for show, or does that really mean something to you?" I welcomed any and all opportunities to share my story and how Jesus had impacted my life. I also respectfully asked people what they believed. I was genuinely interested. I wanted to learn what made the people I work with tick. Mutual respect and understanding one another's story are great places to start any discussion. Arguments and emphasizing differences typically are not.

I realize that American culture has changed since my time at Microsoft. Expressions of religious belief—at least some kinds of religious belief—are not tolerated, much less appreciated, as they once were. Still, I believe the approach of someone who loves Jesus should remain the same. Love is the answer. Love the other person. Share your story. Ask about their story. Listen. Build from there.

- Are you teaching your people in a way that is respectful of their personal learning styles? Perhaps the information you are presenting is correct, yet not having the impact it can enjoy because of the presentation style used. Altering the style of presentation can make a huge difference, as I found out at MSU.

- Embrace the people you work with. Love them as God does. Discover who they are and what makes them tick. Gain permission to share your story. If you want an engaged team, your team first needs to know you care for them each as individuals and also collectively as a team. Working as a team, you can move mountains!

The Power of Two

At this point in the journey, I want to take a moment to acknowledge the superspecial people who worked as my assistants on the Microsoft journey. They were much more than assistants. They were a critical part of the team's success and my success too. They did many things behind the scenes and without fanfare to help us achieve our collective goals. I am forever grateful to Susan (Han) Hammet who was my assistant in Toronto. She and her beautiful family became close friends. Susan died October 31, 2002, of a brain aneurism and will be forever remembered and missed. The other rock stars included Elaine Jordan in LA and Debbie Russell in Redmond, Washington. Thank you!

FIFTEEN

Leaving Millions ON THE TABLE

Susan and I were very blessed financially from my tenure at Microsoft. In fact, financial blessing came early. I became a millionaire at age thirty and a multimillionaire a short time later as Microsoft's stock continued its ascent.

Lessons learned from incredibly brilliant people were another great source of blessing. On many occasions this included Bill Gates himself.

Our team at Microsoft University was number one in the US Sales and Marketing Division's (USSMD) performance to plan. We were the only group to beat plan every month of the fiscal year. I was traveling less than ever and golfing more than ever. Stock options were vesting on a regular schedule and bringing in significant wealth to our family.

Regardless, I didn't feel the support of my direct management. I reported to Bob McDowell. Even though I had top-level performance to plan, I learned I had received the lowest allocation of stock options. Managers were required to allocate their pool of stock option grants across all direct reports. Bob had come to Microsoft from Ernst & Young to lead Microsoft's new consulting business to help our corporate customers implement Microsoft technologies as part of their strategic initiatives. Many E&Y teammates came with Bob. He was the one who rewarded the Microsoft consulting direct reports with the stock options. Even though my team's performance and mine were rated very high, the stock I was allocated was lower, effectively being used to balance Bob's spreadsheet.

Jeff Raikes and I also saw things differently. "I know you will take the hill if I ask you to," Jeff told me one day. "Yet I want you to challenge if we should even take the hill." I understood the point Jeff was making. We just disagreed. Bill, Steve, and Frank's directions about MSU were clear, and the subsequent execution by our MSU business unit exceeded the goals. Still, what we accomplished did not seem valued. When the strategy was to take the hill with Rich and Scott, which I did, I felt valued to the highest levels of the company. But those days had passed. In one year, everyone between Bill Gates and I in my reporting structure had changed. The new leadership didn't communicate value for the work being done by my team and me.

Susan shared with me that, regardless of my performance, I was the unhappiest she had ever seen me. She was right.

I have been passionate about personal information management since the day I put Visidex on my desk at Great-West Life. I kept a list of the features of my ideas for the perfect personal information management (PIM) software tool. Blair Bryant had become a friend when I was at Microsoft in Los Angeles. He had been an executive at DayFlo, the makers of the first PIM software. We talked and compared notes on a regular basis, enhancing the PIM list.

My interest in PIM software resulted in an invitation to a PIM ideation meeting with Bill Gates, Vijay Vashee, and Brian MacDonald. I shared the evolved "Perfect PIM" list as part of that discussion. (Fast forward. Brian was later tapped to lead the team that would create Microsoft Outlook. Microsoft called it a "Desktop Information Manager" (DIM) versus a PIM.)

Many months after that original planning/ideating meeting, I learned that Scott Oki, who had left Microsoft that previous year, was investing in a company called Arabesque Software.

I told Susan that I felt that I had done what I was called to do at Microsoft, but now wondered if it was time to move on and do other things. As reflected above, I was also not feeling valued. I had gone from someone being watched for the potential presidency of Microsoft to a non-priority in the eyes of everyone between Bill Gates and me in the organization chart. Leaving, however, would involve walking away from millions of dollars worth of stock options that were still to vest. Susan and I prayed a great deal about this decision. She saw both my frustration and my excitement in the PIM project. In the end, Susan supported the decision for me to leave Microsoft. I felt nothing but love from her as she supported me in my decision to move on. After several weeks of prayer, discussion, and a feeling of peace, I made my move.

I agreed to sign a non-disclosure agreement with Arabesque Software. Arabesque Software was working on a PIM that came to be known as Ecco. When they showed the software that was still in development to me, my jaw hit the floor. I was looking at the Perfect PIM list in action! I called Blair and told him he needed to get to Bellevue, Washington, as quickly as possible. It was a Wednesday. He arrived the next day and joined Arabesue shortly thereafter. I left Microsoft shortly after that in 1993 to join the Arabesque team. Blair and I were quickly joined by Dhiren Fonseca and Dave Neir. I had first met Dave when I worked with Microsoft Canada; he was in Microsoft's Corporate Finance team at the time.

Financial planners have told me that our next moves would never have been recommended by anyone in their profession. Susan and I were and are okay with that.

When people hear about the time period that I was at Microsoft and look at the stock's meteoric rise during that period, they often assume that Susan and I are independently wealthy. That is not the case. Over the post-Microsoft years, we put aside money to put our kids through school so they could graduate without any student loans. We also kept enough to provide our children with their first vehicle and to purchase a great home for all of us. The rest of the money, which was most of it, we used for others. We invested in software to share the Christian faith and to provide inner-city education scholarships and promote other causes we saw as worthwhile. We were young and felt that if God wanted us to have wealth again, we would have future successes in business that would allow that to be the case.

Quite simply, we did what we felt called to do. Other people may have felt called differently in a similar situation. We simply knew that we could look in the mirror and feel that we had done what we had been led to do.

- Are you following what you are feeling called to do? If not, what is holding you back?

SIXTEEN

Leverage Your Assets: Lessons from Microsoft and *ARABESQUE SOFTWARE*

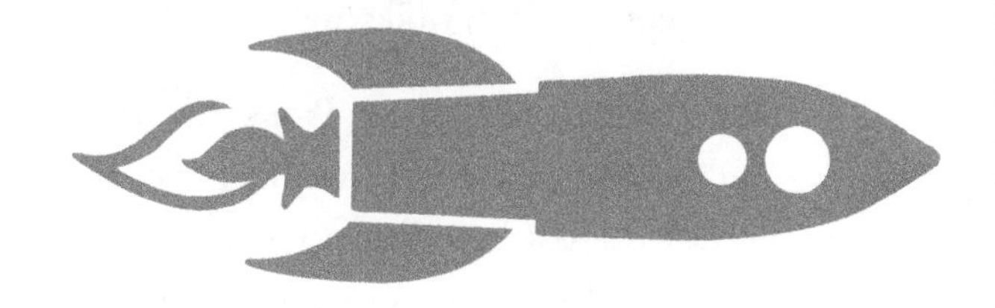

I learned many lessons during my tenure at Microsoft and immediately thereafter at Arabesque. "Pride comes before a fall," unfortunately, would be a repetitive theme.

Shapeware had created a product called Visio. It let the average business user create great charts through a drag-and-drop interface. The team at Shapeware modeled their menus after the Microsoft interface. This allowed them to become a poster child for Microsoft. When Microsoft wanted to show off great Windows applications, they often pointed out Visio. In fact, Microsoft later acquired Visio because the interface work they had done made it feel like a natural fit with the Microsoft application family of products.

Ecco, on the other hand, was led by Pete Polash. Pete had created presentation software called Persuasion and sold it to Aldus. The revenue from Persuasion was the primary funding source that got Arabesque and Ecco off the ground. Pete had gone up against Microsoft before and won. He figured he would do it again. The Arabesque executive team, myself included, encouraged Pete to follow Shapeware's example and make the menu structure common to the Microsoft office application suite menus. He refused. In fact, the Arabesque menus worked differently than almost any other program's menus. As a consequence, Shapeware got acquired, while Arabesque did not.

Brian McDonald brought Project to Microsoft. He went on to lead the development of Microsoft's personal information manager, actually called a desktop information manager and known as Outlook. Outlook gained a super power. It was going to be bundled with Microsoft Office. That meant even if it was average, it would win over other similar available products. By this time, Microsoft Office was large enough to command that kind of power for anything that was bundled into its suite.

The lesson? Leverage your assets. Leverage the leaders in the industry. Play to your strengths.

- Do you have any successes that can be leveraged into new successes in your business? Are there any ways in which you can become a poster child for another entity in your industry to leverage their market position?

Back to Microsoft

Just a few months after we put Arabesque Ecco into the market, I received a call from our representative at Ingram Micro, the largest distributor in the industry. "Congratulations!" the voice on the other end of the phone said. "You're number one!"

What? Already? I thanked the person and immediately went to my teammates. "Bad news. The market is much smaller than all the industry dialogue has suggested to date. We're already number one at Ingram Micro in the personal information product category."

Arabesque was sold to Net Manage in 1994.

I moved forward with my own company, Provident Ventures Inc. My current company, Meta Media Partners LLC, is the successor of Provident Ventures, which was started in April 1994. Microsoft was one of our first clients.

In mid 1994, I returned to work on Microsoft's corporate campus as part of this client work. Rick Thompson, who had led Microsoft's hardware group, including products such as the Microsoft Mouse, was running Microsoft Works. Rick called me and asked if I would join his team as a product manager for the next release of Microsoft Works, which would be titled "Microsoft Works for Windows 95." Rick wanted someone who was technical, who could properly assess the reality of a developer's statements when a certain feature could or could not be done, or when the amount of time estimated to do it seemed incorrect. He also wanted to free two members of his team to work on products for Microsoft Bob.

If you know what Microsoft Bob was, hold your giggles. If you look at Siri and Cortana and other voice systems today, minus the animated Paperclip or Smiley face, you will have to admit that Bob was on the right track. It was just a little ahead of its time. And the animated "Clippy" was not the path that most consumers would embrace as readily as a faceless Siri, Cortana, or Alexa.

Rick asked me to manage half of the features of Microsoft Works for Windows 95, including the spreadsheet charting and ensuring every single Avery label would print and align properly. In fact, the specifications for that release of Works were the top requests that our product support team received for help with the product. We addressed the technical support issues that were creating our users' primary areas of frustration. As a result, the product was well received. Many users marveled at the prioritization of features that had made their way into that product. We simply listened to our customers, and it worked.

The shipping of the Works and subsequent project transition for the next release of the product marked the end of my second tour of duty with Microsoft. I had spent a total of ten years with one of the world's most incredible companies.

Part Three

HOPE FOR THE FUTURE

SEVENTEEN

Losing the Way: *MICROSOFT SUED*

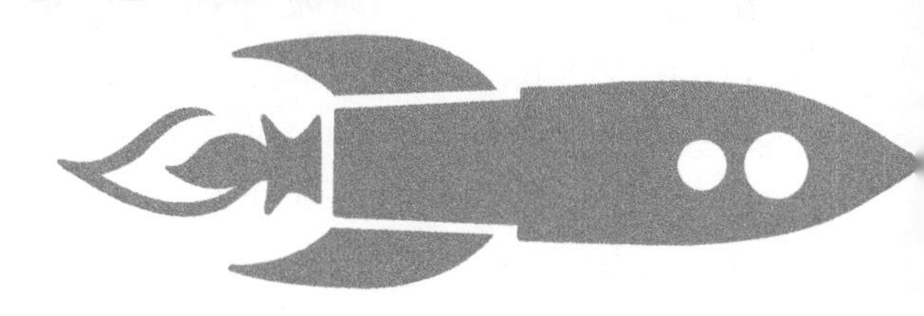

I didn't want to be right.

In the February 1990 management retreat, I cited arrogance as the greatest risk to Microsoft. Arrogance had led to the fall of many great software companies. Unknown to me at the time, starting in 1990, the Federal Trade Commission already had Microsoft in its crosshairs.

Several years later, *Wired* magazine told the tale of the complete government antitrust suit against Microsoft. The article was titled "The Truth, the Whole Truth, and Nothing But the Truth."[2] According to

---

[2] See http://microsoftsecrets.com/wiredantitrust1 and http://archive.wired.com/wired/archive/8.11/. For more on the suit, see the following: http://microsoftsecrets.com/wikiantitrust; http://en.wikipedia.org/wiki/United_States_v._Microsoft_Corp; http://microsoftsecrets.com/wiredantitrust2; http://archive.wired.com/techbiz/it/news/2002/11/35212; http://microsoftsecrets.com/wiredantitrust; and http://archive.wired.com/wired/archive/2.04/gates.html.

Wired's sources, one of the reasons the whole antitrust issue happened was arrogance. Arrogance had opened the door to the antitrust charges.

That being said, the actual antitrust case that US Attorney General Janet Reno brought on behalf of the federal government in 1998 may have been more politically motivated than it was, as claimed, an act seeking to protect consumers. After all, it was Borland, Novell, and other Silicon Valley competitors that brought the charges to the government via Susan Creighton and Gary Reback, lawyers and antitrust specialists with the Silicon Valley law firm Wilson Sonsini Goodrich & Rosati. Consumers didn't bring the charges against Microsoft. Nor were consumers suffering. In fact, software prices had fallen and made computers more affordable for consumers. As I shared earlier, Borland actually brought some of the price pressure on itself by trying to go after Microsoft's applications and then its crown jewels.

Moreover, the timing of the launch of the antitrust part of the government's case against Microsoft has another interesting side. President Bill Clinton, Vice President Al Gore, and Attorney General Janet Reno were under rising public and political pressure at the time. Reno was being pressed to take action against Clinton and Gore for illegally raising funds through a Buddhist temple in their 1992 and 1996 campaigns. News reporters uncovered more and more information that indicted, at a minimum, the vice president. Years of investigation into Microsoft had gone on, with Creighton and Reback continuing to heckle the government to take more aggressive action on behalf of their clients. By moving forward with the antitrust lawsuit on May 18, 1998, Reno effectively shifted the country's focus away from the fundraising scandal and toward the government versus Microsoft. The potential legal quagmire, if not impeachment, for our nation's leaders took a backseat and stayed there.

Microsoft's arrogance certainly helped fuel the Justice Department's fire. And the company's continued arrogance throughout the antitrust suit process did not help either. The Department of Justice investigation, which started on April 21, 1993, continued through the antitrust suit, decisions, appeals, settlements, and extensions all the way through May 2011. This was further extended with the European Union following the US antitrust suits with their own actions, which resulted in over $2 billion in fines and ongoing legal actions through 2012. The US federal government took significant action against Microsoft, first ordering it to be split into two companies—one for operating systems and the other for applications. Through the appeals and settlements process, this was eventually taken off the table. Microsoft had to disclose its APIs and protocols, including internal use APIs not previously made available to any other companies. Microsoft also had to allow a panel full access to its systems, records, and source code for five years.

The consumers hurt most by the government's action were Microsoft shareholders. Over half of Microsoft's stock value was erased. The DOJ had prevailed in decisions that labeled Microsoft's business practices as monopolistic. This resulted in several decisions forced upon Microsoft, including demanding it unbundle its browser, Internet Explorer, from the operating system install process. The battle with the government went on for over half of Microsoft's life to date. In my opinion, the greater loss was heart and energy within the Redmond campus. Bill Gates stepped down as CEO in January 2000, and Steve Ballmer took over the CEO role. Microsoft stock fumbled along under Steve Ballmer. That fumbling resulted in Microsoft stock options becoming essentially worthless to a generation of employees. The "golden handcuffs" were unlocked and significant talent was lost during this period. To Ballmer's credit, he did save the company from being split in two when the DOJ seemed bent on achieving this outcome.

Over the coming years, Apple took over the hearts and minds of consumers. Microsoft was no longer seen as the visionary company it once was. Tech influencer conversations would name Apple, Google, Amazon, and others before the name Microsoft would be mentioned. The company did not start to recover its innovator reputation with consumers until 2014 when Satya Nadella was named Microsoft's CEO.

The Ballmer Years

Microsoft's boisterous leader, Steve Ballmer, was named CEO of the company in January 2000. He remained in that position through 2014. If a for-profit company's role is to provide value to its shareholders, the stock record during Ballmer's tenure demonstrated Microsoft's poor performance. Almost instantly from the time Steve took over, Microsoft's performance flattened. Granted he took over as the US government's rulings against Microsoft were enacted. Still, the company never recovered, as the Microsoft stock performance graph that follows shows.

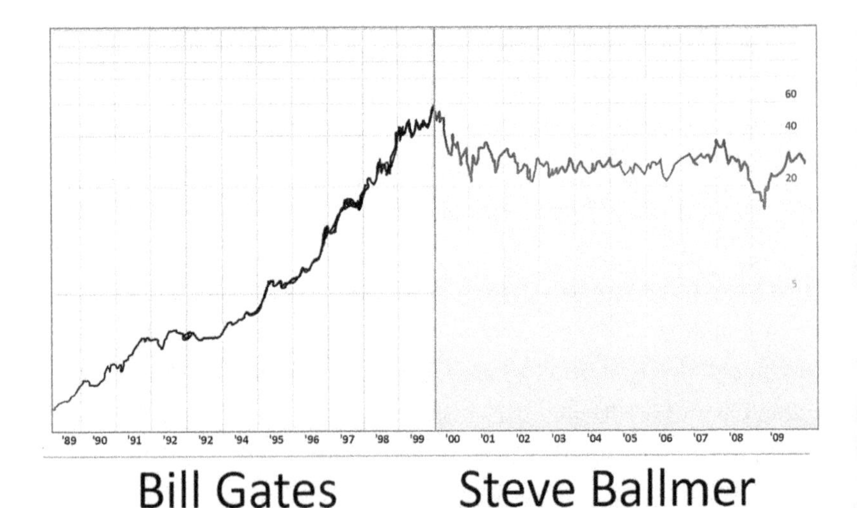

Perhaps the worst reality during Ballmer's reign was the destruction of a country's economy through the moves that resulted in Microsoft's acquisition of Nokia and eventual write-off of the same. Microsoft "wins" at Finland's expense have turned into a total loss as of this writing. Former Microsoft executive Stephen Elop was hired as Nokia's CEO. While at the helm of Nokia, decisions made led the company to significant losses in market share. Nokia had been number one in many cell-phone markets, garnering a 90 to 98 percent share in some countries. Under Elop's leadership, Nokia's revenues fell 40 percent and profits fell 92 percent. Nokia's stock price fell 81 percent over the same two-year period. Elop laid off 11,000 Nokia employees in 2011 and another 10,000 in 2012. On September 3, 2013, the acquisition of Nokia by Microsoft was announced. Microsoft bought the company for $7.2 billion. Nokia's market value had dropped 85 percent since Elop had taken over. Its Smartphone share had dropped to 3.3 percent. Elop returned to Microsoft as executive vice president of the Microsoft Devices Group business unit. He received an 18.8 million euro bonus from Nokia's board. The deal was apparently inked on the day of the sale's announcement to Microsoft. In July 2014, Microsoft announced the layoff of 18,000 people; 12,500 of them were from the Nokia acquisition. On July 8, 2015, Microsoft announced it would layoff another 7,800 people from the Nokia acquisition and would write off $7.6 billion of value from its books. On May 25, 2016, Microsoft announced an additional $950 million in value was reduced from the books for the phone division and an additional 1,850 jobs were reduced.

The bottom line. Microsoft's acquisition of Nokia led to 43,150 jobs lost and the total devaluing of Nokia. This was a completely failed acquisition. Nokia had once enjoyed a 95 percent market share in many markets. Now, it is effectively worth nothing. Actually, less than nothing when one factors in all the related costs of a failed initiative like this.

On top of this, think about all the related companies and their people who supported Nokia in the Finnish economy and around the world. Wiped out.

I stand by the "greatest risk" statement I had made in February 1990. Arrogance was Microsoft's biggest risk, and it led to its fall for many years. That same arrogance decimated an acquired once-thriving company and the home country that once benefitted from its profitability.

As of this writing, the news is now much better for Microsoft. The company's new CEO, Satya Nadella, has brought new hope and new life to the entire company. He was named Microsoft's new CEO on February 4, 2014. That day was an historic turning point for the company. People inside Microsoft feel it, and industry pundits have said as much. I also noticed it the next time I walked on the Microsoft campus.

I received a personal invitation from Rich Kaplin, the new CEO of the Microsoft Alumni Association, to be part of a group of fifteen alumni brought back to Microsoft's campus on February 19, 2015. We met at the Microsoft Executive Briefing Center and were briefed by Windows 10 product management, Microsoft Azure leadership, Microsoft Ventures, and Microsoft Office product management. We were also given a tour of Microsoft's vision for the future and its global security operations. Our opinions were sought and valued throughout the day.

Microsoft has approximately one hundred and twenty thousand employees. It also has more than that number of alumni who have passed through its halls over the years. Over twenty-four thousand of these are active members of Microsoft's Alumni Association. Under Steve Ballmer, the group shared that alumni were considered persona non-grata. *You left, now have a nice life* was the sentiment shared by many with me over the Ballmer years. Insiders expressed to me that Ballmer said as much to alumni he encountered. In stark contrast, Satya Nadella sees alumni as those who should be Microsoft's number one raving fans out in the marketplace. He views alumni as possessing value to contribute to Microsoft's future.

The alumni brought back to the Executive Briefing Center included leaders from all parts of the company during the years of its meteoric rise. Many had not been on campus for fifteen or more years. We shared the perceived difference in the spirit of the company. The presentation teams all saw it and felt it as well.

What is so different about leadership under Nadella? One example concerns the way in which he faced the reality of Microsoft's true place in the market with today's home and corporate consumers. Nadella measured market share across all devices—personal computers, tablets, and phones—to get to Microsoft's true market-share numbers versus solely looking at PCs and even Microsoft applications' presence on the Apple Macintosh.

Also, projects that had been stopped under Ballmer, such as Microsoft Office for the iPad, were given the green light by Nadella and came into the market after he took over as CEO. The former leader of the Office for iPad initiative had left the company when Ballmer refused to release it. Under Nadella, Microsoft has made moves to bring core technologies, including its Cortana voice recognition, to all major platforms, including iOS and Android devices.

Nadella's big move in 2016, acquiring LinkedIn, is a stroke of genius. This move will reap benefits across the board for Microsoft and its business customers.

Like the market-share numbers, Nadella is also measuring Microsoft on the revenue per customer it makes across all platforms. He seems to understand that arrogance should have no place when a company has fallen as far as Microsoft has.

Now is the time for listening to customers. Now is the time to learn from the rocket ride from worst to first and to work to repeat history.

It's a new day at Microsoft.

EIGHTEEN

Vision for Microsoft, Apple, Amazon, Google … *AND YOUR COMPANY*

Competition is great for the marketplace, for consumers, and for innovation. It has made Microsoft hungry again.

In this book, I have shared my story and the stories of others to target, not just Microsoft, but also Apple, Google, and Amazon—in fact, to any company that has risen to a leadership position and to those that want to. Sharing the stories of Microsoft's rise and fall I hope will inspire Microsoft to aim high and will help companies such as Apple, Google, Amazon, and others to aim high as well, including your company.

To become great and stay there, you need to value your people, set a clear vision for your company and your industry, manage with integrity and with principles, know your customers and your competition, be real with yourself and with others, and know the true metrics that reflect your current situation in comparison with where you want to be.

If I was handed the reigns of Microsoft, what would I recommend? Here are my conclusions and recommendations.

1. Know Who You Are.

Is Microsoft still the API company? Not so much. Can it be won back? Perhaps, but only through the hearts and minds of the developers. And today the definition of who is a developer is much broader than it has ever been. Microsoft must make significant efforts to own the hearts and minds of leading software developers on traditional personal computer systems and applications for new mobile and Internet-of-things (iot) devices.

2. Value Your People.

That includes your alumni and customers. Frank Gaudette always said, "People are our greatest asset," and he lived this principle with his actions. Microsoft as a whole and in all its divisions and departments must do the same.

3. Know What Real Value is for Your Customers. Deliver Value.

Microsoft understands this for corporate accounts, but it has almost no clue when it comes to small business and entrepreneurs.

Don't read your own marketing. Talk to the marketplace. In our 2015 February alumni meeting, several alumni pointed out specific examples where common everyday tasks were difficult with Microsoft and its products. Some at the table denied the difficulties, yet numerous facts and examples were cited in rebuttal.

For example, as a test, I took the same approach as my LAN Manager MSU test. I signed up for Microsoft Dynamics Customer Relationship Management (CRM) software to compare it to my experience with Ontraport, Infusionsoft, Emma, and many others. I found that the Microsoft experience was disastrous. The videos are poor. The websites are more confusing than helpful. And even after I closed the trial, I continued to get many weeks of emails warning me that my trial would soon close. How can I believe Microsoft even understands CRM when they do it so poorly in the demo process?

Microsoft needs to integrate with Gmail and other systems. Outlook does a poor job of this. The Mail client in Windows 7 through 10 does a poor job as well. Microsoft needs to use both Outlook and the Mail client to connect with multiple email systems easily. At this time, the two don't integrate well. I could write another book's worth of material on Outlook alone. The Gmail issue also falls on Google for not fully cooperating. Nonetheless, the marketplace views this issue as a Microsoft problem. And, as a result, Gmail is more dominant in many sectors of business than Outlook.

4. Know Your Competition Deeply.

In my experience, most do not really comprehend their competition. Along these lines, you may want to reread the stories about Lotus, Borland, and others provided in this book. You must know your competition so well that you can reverse engineer their strategy.

5. Focus and Leverage Your Business.

Understand how to build focus and leverage into your business. Define how you can create leverage both internally and with your business partners and customers. Focus and leverage create speed and competitive advantage.

For Microsoft, I believe this requires considering breaking itself up into more companies in order for it to grow into the future and become even more profitable. For example, I think Expedia was the right model for Microsoft. Expedia was started in 1996. Bill Gates and the Expedia team spun the company out starting with a 1999 public offering. It was fully spun out as a separate company in 2001 when USA Networks Inc., an IAC company, took a controlling interest in the company in a multi-billion-dollar deal. Expedia grew to become the number one provider of travel services in the world, and it returned great value for its shareholders. Yet Ballmer halted other similar efforts. Now is the time to revisit this strategy.

6. And the Bottom Line: Be Principled in All Aspects of Business.

Principles matter. Let the lessons of the past forty years ring true. Learn from them.

I believe these six priorities can and should be applied to Apple, Google, Amazon, and many other companies. Even if the scale of your company is much smaller than these industry titans, you can apply the six core principles above. It's time to put these principles to work in your business and build your next—or perhaps your first—rocket ride. Even if you are currently the worst, you can rise to be first!

Additional *RESOURCES*

The following resources provide additional perspective and historical value in understanding the leadership of Microsoft and perspectives from this important time period. Additionally, I have provided resources for you and your team to access for developing your team's culture. Plus, I have a few fun bonuses: music, parody songs from Microsoft's past national sales meetings, and more. All these resources are available free at http://MicrosoftSecrets.com/bonus.

- The 1990 Conference Letter and Attendee Roster (included here)
- Chris Peters: Shipping Software On Time (1991/1992) (included here)
- The Microsoft View of the Office of the 90s: Strategy Briefing Tour Presentation, by Mike Maples (April 1989) (included here)
- Excellent Cultures Assessment. (Visit http://MicrosoftSecrets.com/ExcellentCultures)

- Bill Gates explaining the state of technology and the vision for the coming years in a November 1988 address to the Chicago Users Group. (Visit http://MicrosoftSecrets.com/BillGates1998)
- Microsoft mock packaging video, which was created by Microsoft's internal design teams out of frustration, with the goal of motivating the company to simplify its design. It looks at the complex packaging Microsoft creates in comparison to the beauty and simplicity of Apple's packaging. (Visit http://MicrosoftSecrets.com/mock)
- "Grey Flannel Navigator" — A visionary video of where computing was headed, created under the leadership of John Sculley during his tenure at Apple. (Visit http://MicrosoftSecrets.com/greyflannel)
- Microsoft training theme music. (Visit http://MicrosoftSecrets.com/trainingmusic)
- Doug & The Blibbets — live national sales meeting recordings. (Visit http://MicrosoftSecrets.com/Blibbets)

The 1990 Conference Letter and Attendee Roster

Microsoft®

Memorandum

To:	David Jaworski USSMD	cc.	Rich Macintosh Scott Oki
From:	Jon Shirley		
Date:	November 14, 1989		

Dear David,

I'd like to invite you to the first Microsoft Management Conference, to be held from February 23 to 25, at Port Ludlow Resort, about 90 minutes from the Redmond campus.

The purpose of this meeting is to tap a broad, experienced management group to discuss and propose solutions for the strategic and organizational challenges Microsoft will face in the coming years. A handful of people from each of the divisions of the company are being invited. We expect a total of about 35 people to participate, including about six executive staff members. Executive staff members will serve as facilitators and information resources, rather than as participants in the discussions. Our aim is to hear from you and other key managers from all parts of the company.

To kick things off, Bill Gates will give a talk on markets and technology; next, Frank Gaudette will talk about the financial model and performance of Microsoft. The meeting will then break into study groups of about five people each, to discuss and make recommendations on issues of importance to Microsoft. As we get closer to the meeting date, I'll send you a list of the issues we've identified, and ask you to suggest any other issues that could be considered for inclusion in the final list.

Bill and I are looking forward to this meeting. Please let your manager, Susan Raunig and me know if you will be able to attend this important meeting.

MANAGEMENT CONFERENCE ATTENDEES

Mike Appe

Mike Appe is the Regional General Manager for MS USSMD operations in the eastern US. Mike joined MS in June 1987. Prior to this he was Eastern Regional Manager for AST Research, a PC and board-level manufacturer headquartered in Irvine CA. Prior to that he was Assistant VP for Commercial Union Ins., at their US headquarters in Boston MA where he managed their insurance agency telecommunications and marketing efforts. He has a BS in Mathematics (1973) from the University of Vermont.

Adam Bosworth

Adam Bosworth graduated from Harvard in 1976. He built the MIS systems used by Citicorp Retail 1977-81. From there he built the Fund Transfer System at Crocker National Bank from 1981-83. He started the Analytica Corp and built Reflex for the PC 1983-85. After he sold Analytica to Borland and built Quattro for Borland during 1985-88 joined MS in October 1989. Adam is Senior Program Manager in DABU.

David Curtis

David Curtis is a Senior Corporate Attorney responsible for the international group of Microsoft's Dept. of Law and Corporate Affairs. He joined the Company in 1987 after seven years in private practice in Seattle. Dave received his AB in Political Science/Philosophy from Whitman College in 1973 and his JD from Cornell University in 1980. He is also chairman of the Board of Directors of the Business Software Association, a 6 company coalition to combat internatinal software piracy.

Patrick De Smedt

Patrick has been in charge of the Microsoft operations in the Benelux area since 1986. BV is based in the Netherlands and is expected to generate $20M in FY90 with a headcount of 37 people. He joined the Company in 1983 as OEM Sales Manager of MS France. Before joining MS he worked for the European headquarters of Altos Computer Systems as Senior Software Engineer. He has a degree in Commercial Engineering (1977) from the University of Louvain (Belgium).

Rick Devenuti

Rick Devenuti is responsible for the accounting issues rel ating to USSMD, Corpcom, Press, CD ROM and MSJ. Included under the USSMD umbrella are all accounting issues relating to Campus North and cost of goods sold. He joined the Company in December 1987 as USSMD Accounting Manager. Prior to joining MS he spent four and one half years with Deloitte Haskins and Sells, where he was the accounting senior on the Microsoft audit. He is a graduate of the Un. of Washington.

Martin Dunsmuir

Martin Dunsmuir has been Director of Presentation Manager Development since March 1989. He also has responsibility for National Language Support in OS/2. He joined MS in 1985 and was Director of XENIX System V/286 before assuming his present assignment. He graduated from Oxford University with a BA in Physics. Between 1979-1985 he worked for Logica PLC, a large European Systems House. He has been closely involved in UNIX related standards activities for over five years. Martin is the co-author of rtwo popular books on UNIX Programming.

Richard Fade

Richard Fade is currently Director of OEM Sales responsible for US and Canadian OEM business. Richard joined MS in June 1986 as a US OEM Account Manager, and was a OEM Sales Group Manager 1987-89. Prior to joining MS he was a Regional Manager for Triad Systems Corp. a minicomputer company providing hardware and software solutions into vertical markets. He also held customer support, marketing rep. and National Accounts Marketing positions at Triad. Richard has a Bachelor's degree in Business (Accounting and Finance) from Florida State University.

Sam Furukawa

Sam Furukawa has been GM of Microsoft KK since the subsidiary began in 1986. He came from the ASCII Corp. His experiences at MS have included that of programmer, OEM Acct. Manager to promote MS-DOS in Japan, sales and marketing of Multiplan and Chart, to GM of MS Japan. The annual revenue of MS KK in FY8mcame to $45M in OEM and $25M in retail business. The R & D function has more than 50 engineers for DBCS enhancement.

Fred Gray

Fred Gray is responsible for the Microsoft C, FORTRAN, MASM, COBOL, and PASCAL language products. He has been with MS since April 1988. Prior to joining MS he was responsible for the development of Microrim's R:Base DBMS products. Prior to that he managed the development of Boeing's RIM DBMS product. He has also developed software as a member of Boeing's structures research department. He has a BS degree in engineering from MIT (1965).

Janie Guill

Janie is Director of Materials for Manufacturing division. Basically this means coordinating with various groups such as finance, sales, product marketing, user ed and corporate communication to schedule and support the production of MS products. She joined MS in 1985 as Purchasing Manager.She is originally from the East Coast (Virginia) where she directed the Procurement department for a systems house providing computer systrems to the US gov. Her Bachelor's degree is in Business with a concentration in Accounting from George Mason University.

Dave Jaworski

Dave Jaworski is the General Manager of Sales Operations in the US Sales & Marketing Division (USSMD). Dave is responsible for all Inside Sales and Support, Consumer Service, Technical and Sales Training, and all US sales programs implementation. Dave was previously the General Manager of the Western US, USSMD. He joined MS in March 1985 to help open MS's Canadian subsidiary, where he became the National Sales Manager prior to his move to USSMD. Prior to joining MS Dave was the Technical Support Manager for Canada's largest software distributor. He has a degree in Computer Science from the University of Manitoba, Canada.

Lewis Levin

Lewis Levin studied quantitative analysis and finance at the University of Cincinnati. He then attended the Sloan School at MIT where he concentrated on information systems and marketing. Prior to joining MS he worked for Execucom Systems, a mainframe financial planning software company, and Micropro International. At MS he has been product and program manager for Mac Excel, product manager responsible for launching PC Word 4, and group product manager at the Graphics Business Unit. His new positioln is Director of Applications Marketing.

Jeff Lum

Jeff Lum is a Group Sales Manager in Microsoft's US OEM Sales organization, a division responsible for product sales to computer manufacturers in the US. He joined the company
in 1983 as an Account Manager in US OEM Sales. After earning his Bachelor's Degree in Accounting from the University of Santa Clara, he became a Systems Engineer for IBM. Prior to joining MS he was a Marketing Representative for Dun & Bradstreet's Computer Services Division .

Chris Mason

Chris Mason is the Development Manager for Word Processing in the Office Business Unit. He joined MS in 1985, working on Word for the Macintosh (version 3) as a programmer and on versions 3.01. and 4.0 as project lead. Prior to joining MS Chris worked at Chemical Abstracts Service in Columbus for 7 years, programming everything from device drivers to high-end apps on every kind of equipment except Intel machines. He is also the author of Rabbit, the only world-class Rabbit Herd Management and Genetics program. He is a 1978 graduate of Ohio State University.

Richard Mc Aniff

Richard McAniff is responsible for all Program Management for the Network Business Unit. In this role he is responsible for all aspects of LAN Manager product planning, from initial product design and specification to final product delivery. He is also responsible for managing key OEM development partnerships. Before joining MS he was a member of the techinical staff at Sandia National Laboratories, where he managed a group of software developers. He has a Master's Degree in Engineering (1980) from the University of Arizona.

Bob Mc Dowell

Bob Mc Dowell is Vice President of the newly created Consulting division of Microsoft. He comes to MS after six years at Ernst & Young in San Francisco where he was the architect and developer of their office automation. When he left Ernst & Young his practice was the largest of the Big 8 firms that focused on end user computing. Prior to joining E & Y Bob was with the United States Automobile Association where he started as a systems engineer and moved from that to managing a large end user computer project. For eight years he was in the US Air Force. Bob has a B.A. in Economics from the Virginia Military Institute (1968) and an M.S.B.A. from Boston University (1974).

Joe Monteleone

Joseph A. Monteleone as Manager, Corporate Systems, is responsible for several support system operations. These areas include: PCinstall, PCrepair, Corporate LAN, Email systems, FileServers, and International Network Operations. Joe joined MS in 1987 after spending 9 years with AT&T in various Sales and Technical management positions. He has helped design and implement 17 DataCenters for Sears and the financial brokerage firm of AG Becker. Joe has his undergraduate degree in Marketing from Southern Illinios and has done his graduate work in computer science at MIT and UCLA.

David Moore

David Moore is Director of Development for the applications Division and Development Manager for the Entry Business Unit. As Development Manager he is responsible for development of Win Graph, Win DTP, and the PC line of Works products. David joined MS in 1981 as a programmer on Multiplan. He has also worked on PC Word and PC Chart. Prior to joining MS he worked at Boeing as a project lead networking CAD/CAM systems. He was an active member of the IGES standard committee from inception to its acceptance as ANSI standard Y14.26M. David has a degree in Math/CS (1976) from the Un. of Washington.

Mike Murray

Mike Murray joined Microsoft in June 1989 and is general manager of Microsoft's Networking Business Unit. He has P/L responsibility for LAN Manager and SQL server and manages all development, program management, developer relations, user ed and product marketing groups associated with these products. Prior to joining MS he ran a Connecticut-based vertical market systems company which was a subsidiary of Unisys. In the early eighties he put together the marketing team that launched the Macintosh at Apple. He interviewed at MS in 1981 but was rejected because the company already had one non-technical guy and didn't know what they would do with two. He has an MBA (81) and an engineering degree (77) from Stanford.

Dave Neir

Dave Neir is responsible for all Microsoft Corporation's operations in Canada, Australia, South America, Africa, Asia Pacific and the Middle East. He joined MS in 1983 as International Controller and was then promoted to Director of International F & A in 1984. In January 1987 he assumed his current position as Direector of Intercontinental Operations. Prior to joining MS he was Financial Controller for Standard Aero Int'l/Federal Industries Ltd of Canada. He is a Certified Public Accountant by training and experience.

Chris Peters

Chris Peters is responsible for software development in the Analysis Business Unit, whose chief activity is Excel. He joined the Company in 1981 and has worked on a wide variety
of MS products (DOS 2.0, Windows 1.0, Word 1.0, Mouse 1.0 and various versions of Excel). He has a BSEE (1980) and a MSEE (1981) degree from the University of Washington.

Jim Peterson

Jim Peterson is Director of Operations, Product Support Services. He joined MS in 1984 as a systems analyst in MIS. After developing several internal systems he was promoted to manager of systems for the OEM and PSS divisions. He currently directs technical support operations, which has over 300 engineers dedicated to serving MS customers. Prior to joining MS Jim worked for the University of Puget Sound as the systems and programming manager. He received his B.A. from Seattle University and an M.B.A. from the University of
Puget Sound.

Daniel Petre

Daniel Petre is responsible for Microsoft's operations in
Australia and New Zealand covering all OEM and retail sales and support. He joined MS Australia in 1988 in his current role. Prior to joining MS he was Director of Marketing at NEC in Australia. Prior to this he worked for Burroughs in mainframe/minicomputer sales, support and marketing. Daniel has a BSc (Computer Science/Statistics) from the University of New South Wales and an MBA from the University of Sydney.

Greg Riker

Greg Riker is the Director of Development in the Multimedia Systems Group. He is responsible for the development of MS's multimedia system software, including test and turnkey application develeopoment tools. He joined Microsoft in mid 1989. Prior to MS he was Vice President of Technology at Electronic Arts, the leading publisher of floppy disk-based entertainment software in the industry. Prior to entering the computer industry 10 years ago Greg was in the music industry as a recording engineer and record producer. He graduated from Purdue University in 1973 with a degree in Electrical Engineering Technology.

Charles Stevens

Charles Stevens is general manager of the business unit that is developing future database and basic products. He was previously Director of Marketing for Applications and joined MS in 1984 as a product manager in word processing. Prior to that he worked for a short stint in product management at Hewlett Packard and graduated from Harvard Business School. Charles has a BA in English and History from Bristol University.

Patty Stonesifer

Patty Stonesifer is responsible for all editorial, sales and marketing of Press books worldwide. She joined Press in 1988. Prior to joining MS she was responsible for editorial and marketing at the Que Corporation, a computer book publisher and a subsidiary of Macmillan, Inc. She previously
held various training and management positions in large mainframe envionments. At Essex, a subsidiary of United Technologies, she managed an MIS Information Center chartered with implementing end-user computing company-wide. She has a degree in Journalismn from Indiana University.

Marty Taucher

Marty Taucher is Director of Public Relations and Corporate Events. Working with his staff and the Waggener Group, he is responsible for all US Press and industry analysts relations. He also directs much of MS's corporate events marketing activities, including trade shows, seminars and the user group program. Marty graduated from Oregon Statae University in 1977 and has degrees in Technical Journalism and Marketing. He joined MS in 1984 and has worked for two leading Northwest high-technology firms, Tektronix and John Fluke Manufacturing.

Greg Tibbetts

Greg Tibbetts is responsible for the localization and production of Microsoft OEM and Retail products in all languages (except Far East). He first worked for the Company from 1980 - 1982 as Manager of Technical Support (PSS), rejoined in 1984 in OS/2 development and moved to International in 1987. In the interim, he was VP Software
Development and VP Engineering for two Microcomputer companies, Lobo Systems and Rana Systems, in Southern California. Prior to 1980 he spent 8 years holding various positions in HR management for Atlas Foundry and Machine Co. He holds a Bachelor's Degree from The Evergreen State College, with majors in Mathematics and Business Administration.

Joe Vetter

Joe is responsible for Reseller, Distribution, and Corporate sales for the Southwest US including CA, AZ, NM, CO, UT, NV and HA. He joined the company in 1984 as an OEM Account Manager responsible for the company's single largest revenue source, Tandy Corporation. Prior to joining MS he worked as a Design Engineer for Chevron USA. He holds a Bachelors degree in Electrical Engineering from the University of Washington.

Interoffice Memo

To: Management Conference attendees
Executive Staff Participants

From: Jeremy Butler
Date: February 12, 1990

Please join Jon and me in welcoming Phil Barrett to the Management Conference group. Enclosed you will find his resume as well as an updated list of those who will be attending.

We look forward to seeing all of you on the 23rd.

Jeremy

About Phil:

He graduated from Rutgers in 1975 with an AB in Mathematics. Following that he was awarded an MS in Computer Science from the University of Wisconsin. Currently he is the development manager of the DOS/Windows Business Unit. Prior to joining MS he held several SW development positions at Intel Corp. He was the Unix 386 project manager and architect, Xenix 286 project manager, Xenix 286 project leader, iRMX 86 project leader, iRMX 86 SW Developer.

Chris Peters: Shipping Software On Time (1991/1992)

Shipping Software On Time

Chris Peters

Chris Peters has been at Microsoft since December, 1981. The following is a compilation of notes taken by individuals who attended Chris' Tech Talk to developers in January, 1991, and a similar presentation repeated to program managers in October, 1992. The videotape of the 1/91 presentation is available to check out through the MS Corporate Library, or order tape #237 by emailing *msstudio*.

The case study discussed in this presentation is Microsoft Excel 3.0 for Windows, which shipped just 11 days later than scheduled one year prior. At that time, Chris was the Development Lead for the project.

Role of individuals within business units

Everyone in the business unit has the same job. That job is to SHIP PRODUCTS. It doesn't matter if you are in development, test, program management, user education, or product management; your job is not to code, not to test, not to manage, not to go to meetings. The job description of everyone is the same "ship products." An expansion of this statement is "ship products EXACTLY ON TIME." Every day you should do what you can to maximize that goal.

Desirability of fixed ship dates

Eliminate the term "target ship date." It implies no commitment. It undermines the central goal we should all have, which is to ship a quality product on time. There are many variables in a project schedule which *should* be variable but the ship date is not one of them.

Fixed ship dates:

- Force creativity
- Force decision making
- Require complete and total commitment by the entire business unit

Work towards fixed schedule dates in the following way:

- Don't try to have a schedule until you have a spec
- Get line individuals to estimate their own schedules so they have higher commitment to making the dates
- Add buffer time to account for the unexpected
- Set 3-month milestones, each with buffer time
- Get everyone to buy into the fixed date and milestone dates up front

Don't let features creep in There will always be another release. You should already be convinced you've got a good product. If the feature is really desirable, trade it for one you've already got planned. The difference between twenty new features and twenty-one new features isn't going to make or break your product, but it will make you miss your date. Will the product flop without the twenty-first feature? If not, do not slip the schedule in order to get it in.

Set milestones Big milestones make the date seem closer and give opportunities to adjust the schedule. People can be motivated to meet a date eight weeks out much better than fifty-two weeks. Don't bother to adjust the schedule until the milestone is completed.

The importance of clear goals

Keep the goals simple. By concentrating on a date, getting everyone in the business unit to be committed to the date, and making that the essential goal, everyone is striving for the same thing. Clear goals include what the product is and, also, what the product is NOT. Simple goals help in communication.

For Excel 3.0, the goal was "sexy, cool features with an emphasis on graphics." The Excel team wrestled with the idea of putting 3D spreadsheets into the product. This was a difficult decision, since Lotus was using this feature as a key differentiation point. The answer was ultimately "no" for a simple reason: added 3D did not help the goal. If the goal said "great analytical capabilities with an emphasis on consolidation" then 3D would clearly have been in, and many features that ended up in Excel 3.0 would clearly have been out. The whole team must have clear goals to ship software on time.

The usefulness of product specifications

It's a common myth that all problems with the development process are due to incomplete specs. Specs are not perfect and will always be incomplete. A spec cannot be "complete," as the competition is always variable. As an agreed upon list of tasks, specs give structure to the project. Specs represent a common theme so that the product looks like it was written by one person. It's everyone's responsibility to create and maintain this common theme.

Developers need to know the spec for a feature, and to know how the competition has implemented that feature. Developers should use creativity, innovation and their understanding of the competition, within the bounds of the spec and with respect to the competition (never forgetting that their number one goal is to ship a product). This is the most productive way to proceed. It avoids the Program Manager having to divine a perfect feature through many iterations of spec generation with long critiques. The team ends up developing the features of a product together.

The point is to strive for the proper level of detail in the spec, while maximizing the innovation in development. Scheduling begins when the spec is "complete." Complete means when every major feature of the product is described in enough detail so development can figure out how to implement the product. This does not require that every bit in the UI be defined. Program managers won't always be able to determine which feature area has major development ramifications and which don't. This is another good reason for the team to work closely together in specing the product.

Constant nature of forces which delay ship dates

At some point during the project, all team members will wake up in the middle of the night positive that it is everyone else's job to make sure you ship late. This includes International, Program Managers, Marketing, User Education, Executive Management, etc. Chris refers to this as the "critical feature game." Everyone has the one "killer" feature that if you don't add, you'll fail. This is not true. Even if you add *that* feature, there will always be a new "killer" feature not in the product. You have to stick to your guns. Or, creatively figure out a way to make it happen. For example, if one of the Product Managers on the team knows how to code, pull him/her off a marketing project to write the desired feature. (Yes, this has happened at Microsoft.)

Tell executive management "no." This doesn't always work, but the more often you can convince them not to add features, the more likely you'll ship on time. Management will accept "no" for an answer, and respect it, provided you are super-prepared and have critical developers on hand with all the answers ready. Don't EVER be unprepared.

Scheduling for an uncertain future

A date more than a year in advance is only a guess. The best we can do is a year or eighteen months; the future is too uncertain to honestly target a date further out than that and make it. With a one year time frame, uncertainty is decreased. You have a reasonable idea what the competition is up to and what customers want in a year. Excel 3.0 set a fixed ship date about one year into the future. They missed this date by 11 days and this slip was a hard decision made near the end of the development cycle.

They were ready to set a fixed ship date after program management "finished" their specs; development, user education, and testing did their schedules; and after everyone had input and agreed upon the date.

Since no one can have perfect knowledge of the future, always add buffer time. For Excel 3.0, 20 percent of the schedule was buffer time. The schedule was made assuming 22 developers working full-time for a year. In reality, the effort peaked at 34 developers. (Adding this number of developers also reflected the commitment of the business unit to the product ship date.) Doubling the estimates does not work, because people will just meet the doubled schedule.

Dual role of team schedules

How do you set a schedule that is both *motivational* and *realistic?* Have each developer create his or her own estimates. This schedule is probably optimistic. To ensure it's also realistic, clarify that the estimate should include non-programming time (attending meetings, interviewing, eating lunch). For example, if you say it's going to take three days to code a feature, that should not mean that you're heads-down writing code for 10 hours a day, with no interruptions. The 3-day estimate should factor in non-programming time. However, this time should not be used as an excuse that the feature didn't get coded in 3 days.

Use of milestones for measurement and motivation

Milestones divide the project into small pieces (3 for Excel 3.0). At each milestone you pretend to ship the product. You fix bugs and test until development reaches an inefficiency point (not enough work to do). The team rallies around each milestone as a real ship date. Everyone talks about the next milestone a couple months away, not the ship date which is a year away. "Only 40 working days until milestone ..." "Three more weeks of debugging..." Keep in mind, however, you don't SHIP a milestone. For each milestone, have a post mortem where you make adjustments in the project (you cannot adjust the fixed ship date though). Milestones, unlike ship dates are not set in stone. For Excel, they knew they were ready to move on when the development and test team's time wasn't productive enough. They also did a mini post-mortem at this point to make sure they were going in the right direction.

For Excel, each milestone was met, but 3 weeks were used from the 9 week buffer after each milestone to adjust for the uncertain future. Features were also moved from group to group at the end of milestones in order to load balance the team and get the product done as quickly as possible. Do not use buffer time to implement the twenty first feature. Buffer time is there to account for optimistic scheduling and the uncertain future

Proper Relationship Between Testing and Development

Unfortunately, it's easy to form an adversarial relationship between Development and Testing. These two should be married. Developers should never resolve a bug as "not reproducible" because that implies that they think the tester was hallucinating in their office. You should talk to the tester and figure out the problem. Don't throw weekly releases over the wall to test. Continually work with testers on a daily basis.

Give private releases to a tester before code is checked in to get most bugs out before everyone gets new code. Run quick tests every day to ensure prior functionality isn't lost and to control regressions. Build often and test the overall functionality of the completely built product - it is possible for each piece to work but the product as a whole to be broken. Provide minor humiliations for people who check in stuff which fails the quicktests. (In Excel 3.0, everyone got a lollipop who messed up and had to tape the "sucker" wrapper to his relight). Assign testers to development teams. You can't ship a product without the testers, and you can't ship it without the developers.

Working with a large development team

First, it DOES work. Everyone should take ownership in shipping the product. It is absolutely essential that you move responsibility very low in the organization. Break the group into feature teams. Each team has its own schedule and Program Manager. For Excel 3.0 there was no integrated development schedule. There were 5 separate schedules run by individual teams. The testing group was organized in parallel with feature teams supporting each of the development teams.

The key to working in a large development team is to keep the message simple. Meetings and interviews should be kept to a minimum. Remember your goal is to ship products. Adding people to a project is an easy method for gaining productivity, but not at a 1:1 ratio. For Excel 3.0, they had 34 developers over the course of the project. This provided only twice the productivity that 10 developers would have given, but that was still a great competitive advantage in producing a timely product.

Zero defects and other methods

Don't confuse the method with the master. The master is shipping on time. The method is zero defects or something else. The method may help achieve the master. Excel 3.0 used zero defects minimally. They had a quick test that ran for about 15 minutes on a daily basis and validated that the core functionality was always intact. If the quick test failed, all development stopped and the problem was fixed. The quick test was simple to run and totally automated. The group assistant started it everyday. (*Note*: don't get caught up in writing automated feature test suites only for the sake of being automated. Remember the goal is to ship products on time.)

The quicktest was created by development, but implemented by testing who could change it at will. Developers were required to check in all code by 2pm. The builds started at 2pm and finished at 5pm, when quick test was run. If one feature team got to zero defects, then they began helping another team fix bugs until they were at zero defects.

Final thoughts

- Use a fixed ship date with everyone's buy-in
- Use milestones
- Plan for the near future (1.5 years max)
- Excel 3.0 was successful because it NEVER occurred to anyone that the ship date would/could slip
- Large teams can be successful

Ways to improve

- Do more code reviews.
- Use more detailed automated tests specific to each feature team.
- Never sign off on disks after 6pm -- you're too tired to make rational decisions.

Q & A

How do you schedule for internal dependencies?

Allow 3-4 months between the ship date for the product you are dependent upon, and yours.

When do you switch to new technology vs. continually updating existing code?

There is a 20% "tax" (fixing most damaging bugs). A new operating system is a good time for a rewrite.

Example schedule

Assumptions:

- Project length - 80 weeks
- Estimate 7-9 weeks coding/3-5 weeks debugging for every milestone. Longer makes the milestone too far away for the team to get psyched to make it. Shorter causes too much overhead.
- First Zero Bug Release candidate = 6 weeks from RTM
- Release Candidate 1 = 2 weeks from RTM

8 Wks	9 Wks	4 Wks		9 Wks	4 Wks		9 Wks	4 Wks		9 Wks	6 Wks		18 Wks
Initial Scheduling Writing Spec Decom-pressing & Recom-pressing	Code	Debug	MILESTONE 1	Code	Debug	MILESTONE 2	Code	Debug	MILESTONE 3	Code	Buffer for Uncertain Future	CODE COMPLETE	Final Debug and Test

The Microsoft View of the Office of the 90s Strategy Briefing Tour Presentation

By Mike Maples (April 1989)

I'm in the area of applications development at Microsoft. At Microsoft we have two divisions -- a division that builds systems and a division that builds applications. And the primary difference between a systems product and an applications product is that an applications product is something that users use; in other words they sit down and use a keyboard or mouse to do something. And a systems product is something that an application uses. I will use a data manager; I will use a communications link; I will use an operating system or a graphical user interface that the system has, to present the applications to the users. In the presentation I'm going to try to profile the entire company. I feel relatively comfortable talking about either the systems or the applications products.

In terms of availability of things, our development is still somewhat of an art, and we schedule very aggressively and seldom do as well as we would like on our schedules. But what I'll try to do is give you a rough frame. I'll use the terminology of a short time if its within the next 6 months that the things we will be talking about will be shipping. I'll give you a time frame of 6 to 18 months for medium -range, and after 18 months for long range.

Let me get a profile of a little of what you do, so I'll understand the level of knowledge and expertise you have. How many of you have some kind of workstation on your desk - PC or Mac, etc? How many would consider themselves an advanced user of the PC? How many people use Macintoshes? How many would say their company is predominantly PC - 60-70% are PCs versus Macs? How many would say that of the purchases they're making today, 40% or more of new purchases are Macs?

Let me go through a set of slides that will act as a reasonably high level overview of what Microsoft is trying to do, who we're trying to do it for, and what we think is important. And then we'll go through demonstrations of some products that we're going to release shortly, or that are currently released; tell you about what we think is important in those and why that will be something you'll be interested in the future; and then spend time on whatever questions you might have.

This is an outline of the slides that we'll go through, and the first thing I'd like to talk about is why we think the world's changing, and what it's going to look like when it changes. We'll talk about how we see personal computers being used through the 90's. And then talk about the parts that we're building for this in both systems and applications.

In terms of what's causing it all to happen, I always like to start back with the technology. It's not that you want to deal with technology for technology's sake, but Microsoft as a company starts off with a view of life that technology allows change for the good -- for the good of the business, for the good of the individual. And our job is to try to implement that change, or implement that set of capabilities into products that you can use. The current set of products that we have, started with a processor chip. It really is a marvelous technology innovation that has had such a dramatic impact on our lives. When you start with the chips, sometimes all you do is focus on the chips, but it's not only the chips, it's virtually everything that's in the system that's changing at a rapid rate. No matter how you measure it, if you talk about capacities, or you talk about performance, they're always up dramatically. If you talk about prices per unit of anything, they are down dramatically. It's not only the chip, it's the memory, it's whatever.

The way I try to keep this in perspective is that I try to say, well how has it been changing? And is it going to continue to change? The scientists tell us, whether it's IBM or AT&T or the Japanese or whoever, that the rate of change from now through the middle of the 90 decade, through 1995 at least, will continue at at least the same rate. Potentially it could change at a faster rate than it's changed to date.

Now let's just put that in perspective. Late in the year in 1981 IBM announced the original PC. In 1984 Apple announced the Macintosh. If you remember, the original PC had a standard memory of 16K and you could go all the way to 64K on the planer. A big system in the early days was 128, and when you had 256, you had arrived. The original system had a single 160KB floppy and you could go up to two; there weren't any hard files available. There were eleven rip-roarin' applications, and I imagine if I gave you a test not one of you could name one of them. Things like Easy Writer, and things that have just gone away.

So if you think about the average user buying 64K systems then, today I would venture to say you wouldn't buy any systems that aren't at least 640K. I would doubt that if you looked at a 160 or maybe 260KB drive, that you wouldn't buy anything today that didn't have a hard file, maybe 20, 30 or 40MB of memory. So it's not the extremes that have gone up, it's the averages. And they've gone up by factors of ten or more.

So let's just think now, what's a system going to look like in 1994, in that same kind of time horizon? Would there be any reason to believe that the average system that you buy would maybe be 16MB of memory, maybe have devices like CD-ROM or fax or image or things that you don't even know about today. So if you keep that perspective of change then you can begin to see what we're trying to put in these systems and why. But change for change sake is not a reason that you'd want to invest or that we'd want to invest. We have to figure out what we'd want to do with these systems and why we'd want them.

So let's talk a minute about the three uses of the computers in the 90's. I don't think there's anything revolutionary here, but I just want to try to structure this, and at the end talk about the products and how Microsoft wants to participate in these three opportunity areas.

First is individual productivity. It's the kind of thing that started this whole revolution. It was the guy sneaking the computers in to do the work that they wanted to do, when they wanted to do it. A lot of us even fought that, but I think that the masses have established the value of that to some extent. And we've all pretty well embraced that technology. American business as a whole has bought 20 to 30 million of these kind of things.

But there is still a lot to be done and what I hope to show you in the demonstration period is to show you some of the technologies and some of the things we can put into these systems that will change the level of productivity that an individual can get out of the system. You can get a preview to some of these things just from the Macintosh. Some of you use Macintoshes, and what I find in a lot of places that I visit is that more and more users are insisting that they get Macintoshes instead of PCs. What they're really buying is a set of features and capabilities generally classified at graphical user interface, ease of use, ease of learning, and so forth. And while you can do a lot of those on PCs, a lot of companies don't allow them to be done on PCs. So the way the department manager or the individual or whomever has the requesting authority gets them these features is to say, "I insist that I need a Macintosh for these set of reasons". And that way you can't take away that from them. There's no way you can't let them have the graphical user interface and the ability to deal with the system more like the way they want to deal with the system.

There are a number of things that allow this quantum change in personal productivity. It's the ability of applications to share data; not only with each other but among themselves or among users. It's a concurrency. It's the ability to have applications being available for your use when you want them. I have the sense that these systems are more acceptable when they model what you do. And if you think about what you do - if you're sitting at your desk and you're reading a report, and the phone rings; very few of you would take the report, fold it up, put it in the file folder, file it in your desk, and turn around and answer the phone. But that's kind of what you ask a personal computer user to do that's running an application. Most of your users today would be asked because of memory limitations or because of the operating system limitations or something else, to stop the spreadsheet they're working on, and initiate a new program if they really had to look in a database. So the concept of the system paralleling the worker's activities is a very important concept in personal productivity.

And shared data - this is an area, group productivity, that is getting a lot of press today. Everybody likes to define it from being anything from one user talking to another user, all the way to the full corporation's business. Let me define it my way, and at least for the rest of this morning, I'll play like I have the definitional authority.

Group activities by my definition is the activities that the group chooses to do to enhance their group's personal productivity. It's not the corporate business. So what that really means is that it is doing for small groups who have discretion over how they attack a problem much like the personal computer did, for the individual who had discretion over how to attack a problem.

What are the things that groups need to do? The first thing is that people in groups communicate. For personal productivity applications may communicate for the benefit of the user, but in the case of groups, the individuals communicate one with the other. The second thing that groups do is that whatever activities they work on, often the work has to be shared among them. You either have reports that somebody prepares and they're shipped to somebody else to make comments on, to make suggestions for improvements. You may be writing a proposal or your monthly activity report where each department manager writes a section and you consolidate or merge them together. You may have spreadsheets that do that. There are a whole set of activities that are really extensions to personal productivity that are really just features on personal productivity products that allow multiple people to use the same product. The third area is coordination. Coordination is something that groups have to do that individuals don't. Groups have to know how to schedule meetings, how to keep a calendar. You keep information about what you want to coordinate or what you want to do. In fact you can go so far not only to keep information, you can manage the activity or the coordination. There's this whole class of software called Nazi software, where you make commitments, one to another, and the software finds out if you kept your commitment. You promised to have this done on Friday, and if you don't get it done today you know what's going to happen . . . So it's the whole concept of management and coordination and commitments that can be assisted and dealt with from groups.

And last, but not least, is work flow. The example I used on the slide is an easy one to relate to. It has to do with if you finish this meeting and you go back and want to submit your expense account, and you say I went to breakfast with Microsoft and I had to spend $8.00 to park. And so you write that down and put it in your Out basket. And your secretary will pick that up, maybe fill out another form, maybe look at it, maybe do something with it. Route it through a mail system – maybe somebody with a little cart comes by and picks it up, carries it to your supervisor or your manager. They'll review it, sign it, put it in another Out basket which will go to another secretary to another cart, and then go to the accounting department where hopefully they'll cut a check, send you the information, and update the business records that there's been some business expense.

Now that's not too hard to model in a computer system. But what happens if your manager decides that $8.00 is too much for parking, and then he kind of reverses the flow. And so the work flow system has to be able to accommodate exceptions. Well what happens if three weeks later you haven't got your eight bucks and you can't go to lunch. It might be that you'd like to call somebody and say, "Where's my eight bucks?". So today you call you secretary, your secretary calls your boss's secretary, saying, "Did he sign it? How come he didn't sign it - is it still in his In basket?" And she says, "No, he signed it a week ago", and so you call and raise cane with accounting saying, "Where's my eight bucks?" But there's this whole process of being able to dial to the processor or the work flow and figure out where a product is. All of those things have to be done to replace a lot of the activities that groups do today.

The last area is the business of the business. And it really is the most important area for the use of these machines. Whatever business you're in, there's a certain set of things you accomplish that are driven by the business needs. It's Delta Airlines who are replacing their reservation terminals with PCs. And they find that they can reduce the key strokes in making a reservation from 280 down to 70. And you can measure that as productivity or less reservation clerks or better service. But the fact is that in a very short time they can justify a four or five thousand dollar list price product to replace a thousand dollar list price product. It's that kind of productivity that's applying this intelligence and this logic right at the end user's area, that allows these systems to grow into doing those activities too.

Now if you believe consultants and you read their studies, you might have read a study done by Nolan Norton and Company - it's a consulting firm out of Boston. They did a study on the use of personal computing and the justification and the cost. And their study says a couple of interesting things relative to this meeting. First, that the cost of owning a PC is about four times the cost of the hardware and the software. So if you're going to buy a computer, and you're going to spend about $5,000 for hardware and software, you ought to plan on spending about $20,000+ on the ownership of that. The second thing that they said is that this type of activity, using the computer to address real business activities, is really the greatest payback. In fact they projected that using personal computers in this line of business activity had a payback of over ten times return on investment, and it was by far the largest of the returns. And it's enabled by the kinds of things you obviously know in terms of sharing data, and tying into the corporate networks and so forth.

I think the really important thing about these three areas, is not that there are three areas, and not that most personal computers today are doing one of them. And in some cases you use terminal emulation, or some other kind of a capability to try to do a hot key between and try to do two of them. But I think the thing that is really important, or the change that we will see between now and the next six or seven years, over the strategic period, is the fact that most individuals will be doing all three concurrently on the same work station. That has some very interesting ramifications in how you organize the ownership or the control or the management. Because you really now have to deal with the fact that both the user has some control, and the corporation has some control and some management responsibility. It will be a more complex undertaking to deal with.

Let's talk a little bit about the Microsoft solution and how the pieces fit together. I break this into four discussions. The first is to talk about what is a workstation and what's in a workstation. I only do that because Microsoft is really committed to the concept that every single person should have a personal computer on their desk, independent of any kind of job classification they have. And at that level of productivity and function will benefit the individual and corporation. Behind or inside the personal computer is a set of sophisticated software to control it. We'll talk a little bit about that. The systems then need to be interconnected. I'll spend a minute or two on that. And last, but not least, they run applications.

So this is going to be a 900,000-foot overview of the subject. This is meant to be a personal computer. I'm not too good an artist so I draw circles and squares. But what it does do, is that it talks about what are the elements that will be available to a user - either in their personal computer, or available in their local area network. These elements are applications and development tools, file systems, operating systems, user interface, and communications. I have arranged them in the circle very carefully. And I believe that any one of them only talks to the boundaries that they show in the circle. In other words I don't believe that many application programs will talk to communications. Application programs will talk to file managers, to operating systems, and to user interface constructs. To the extent that there is any communication going on, it's one of those subsystems that's asking for the communication. It's distributed file management, it's distributed operating system facilities, or distributed user interface and presentation.

At Microsoft in our applications area we target three environments: the Macintosh, OS/2 with Presentation Manager, and MS-DOS with Windows. And what I've done on this slide is taken areas and greyed them out for the places that Microsoft does not participate. Any place that has color, we are actively developing software that can be used in those platforms to deal with business problems. Now the reason that we choose not to do some of the areas is that there are adequate capabilities there now. Secondly, sometimes vendors have a set of beliefs that they can control that better than having a partner work on it. And thirdly, there are cases like an advanced file system in DOS where the machine or the operating system doesn't have the capabilities, the memory, and the facilities to deal with it properly. And so we just choose to say that that's a limitation of the software and not necessarily what you would like to have.

Let's just focus on the three in the middle: the user interface, the operating system, and data management. I will use OS/2 as an example. It probably isn't important which operating system I use as an example. I'll only use OS/2 because it has the most capabilities and the most advanced set of things. And what I'm really focussed on is the set of things that application developers need and users need, to control the system. So I'm not doing it because it's esoteric or because the system does it, but that it's the kind of a thing that as an application developer, or what an end user wants to do through an application, that needs to be there.

In terms of the major part of the operating system, or the kernel - there are really five categories of system facilities, and the operating system needs to own and manage them all. It needs to do multi-processing, multitasking, and have the capability of protecting one application and one user activity from another through some kind of storage protection and security mechanisms. We can spend hours on the subject, but I think you probably have a great appreciation for what kind of capabilities an operating system needs to have.

In terms of a user interface, I think it's really important that we have a user interface that has three characteristics. One is that it's relatively fun to use. That's important. Mac people have established that Mac users like user systems more than PC users do. It means something to have it be pretty and nice. And we need to do that. It costs nothing, and we ought to do it. Secondly, certainly the most important, is that the user interface needs to be consistent. The PC allowed developers to have a great deal of freedom, and they chose to do a lot of things in a lot of different ways. To the end user it often shows up as inconsistency, and what happens in inconsistency is that you never are comfortable that you know what to do. For example if you use one of the old generation spreadsheets you type a slash-F-S and you'd save the worksheet. If you happen to use an old generation word processor you could type a dot-F-S, and it wouldn't save the document you created. It would erase it.

Now what do you think happens when a user does one of those at the wrong time? They'd probably be pretty uncomfortable about doing it at any time. And so now they're very dependent on reference cards or helps or something to do the most fundamental thing as part of their operation. So their learning is not synergistic and it doesn't build upon itself. It builds a wall that they know they don't know what to do next, or they're afraid of what to do next. So the consistency in how applications are implemented is a really key feature. And we're going to show you that across a broad range of applications.

And third is the graphical user interface and some of the things that it brings. Besides ease of use and friendliness, one of the things that a graphical user interface does is it isolates the developer from the device characteristics. Two guys wrote the first release of Lotus 1-2-3 in nine months. A few months later, they had a second release of Lotus. For the next three years they grew their development staff and the majority of what they worked on were printer drivers and screen drivers and plotter drivers and I/O device support. The industry was exploding with the number of screens and the resolutions and the printers. Lotus has over a hundred printer drivers. But it wouldn't be so bad if only Lotus was doing that. Microsoft Word has 147 printer drivers and over 30 screen drivers. WordPerfect, you name the application, all of the investment that the industry was making over the period of three years in the middle 80's was going to building device driver support.

That's not a very good use of creative talent when there are other problems to be solved. So what happens with a graphical user interface is that somebody builds a device drive once, and you install it in the operating system. Now hopefully the world will come to the point where that ought to be built by the guy who builds the hardware. There's a lot of good reasons for that. The main reason is that the guy who built the hardware has all the reasons in the world to make it as efficient and as effective, and to support the device in the best way he knows how, because it's his device that he's getting revenue for. So I think in the long term when you buy a device, a new screen, or a new printer, or a new whatever, a device driver will come with it, just like a cable comes with it, that will install in the major operating systems.

The second thing that this device independence does is it frees the dependencies of hardware and software one on the other, for you to use or take advantage of technology changes. If somebody tomorrow invents a new whizzy great screen that has a whole lot of capabilities that you want and can afford, all you have to do is get one device driver, and every piece of software that you have that uses the graphical user interface will run on that device. So the device frees up the interlocking - e.g. "I can't buy this new printer, because Lotus doesn't support it yet, or because WordPerfect doesn't allow me to use the plotter yet", and so you've been in this constrained environment trying to keep track of all this information. And that can be solved by the graphical user interface.

The third area is advanced data management. Microsoft is of the position and view that it is really an important area. In the next strategic period the vast majority of data will be stored in a relational database and a user needs to have access to a relational database, either on their workstation or on a server or tied into a host system, be that a DEC system or an IBM system, or whatever kind of a system you have.

In addition to having an advanced relational database, it's really important that we do a lot of work in just the file system itself. What we've done is that we've tied the definition of what a file is and contains to eight characters, a dot, and three characters in its file name. We do keep a little bit of information like the last date it was used and its size. But there's a great deal of information about files that ought to be kept. And it really ought to be kept by the system. Things like who created it, when they created it, what application they used to create it, what applications is it linked to, who depends on it, when was it last updated, and what is its security level. There are a large number of things that ought to be available to applications and to users about the files in their system. As the disk devices get larger, as you start attaching to file systems, and as you start attaching to hosts, the ability to find the file among the thousands that you haven't used in a period of time, gets to be a harder and harder problem. And it really is an operating system problem, not an application problem.

If you are a user of Word you probably know that we have the capability to keep information about the user and when it was created. And at least for Word documents you can go back and search on subjects, and on text, and on author and so forth. And that's probably more than adequate for the time being. But in the long term, a lot of these documents are not going to be used by just the application that created them; in fact we're going to show you a lot of cases where an application creates a set of information then it's linked to a different application. And now we want to make sure that when we delete that information, that both applications know about it and have the chance the say it's OK to delete it. All those things need to be built into a file system.

In terms of communications I'm sure you're aware that the ability to tie the systems together to do emulation, resource sharing, communicate one workstation to another, and that really boils down to an architecture that we believe is not arbitrary. It is the architecture we believe most users will have in the 1990 time frame. We call it three-tier. Some people call it two-tier. Some people might not even call it a tier, but essentially it's make up of a series of clients tied over a local area network to a server of some kind. On this slide I happened to choose OS/2 as a server. This might be a mini, or even a mainframe host. This acts as a gateway to the communications vehicle to both hosts that you may own or hosts that you may get services from. These clients can use presentation services, applications file services off of the server systems. I don't think that there is anything revolutionary in any of these topics. I just wanted to do it to show you the big picture of applications.

What I've talked about so far is the system facilities that are in the operating system. That happens to be everything above the red in these systems. In the case of the pieces that Microsoft works on and develops, you can't buy those from Microsoft. We build those as a wholesaler and we sell those to hardware vendors and software vendors who retail and market those to you. So while we work on them and make them available, you don't buy LAN Manager from Microsoft, you don't buy SQL Server from Microsoft, you don't buy OS/2 and DOS from Microsoft. We think it's really important that we work on them though, because the applications that you do use and the vision of how we see this industry growing, and the use of these computers is often dependent upon having the right set of facilities available so we want to make sure that either the vendor's focused on it or that we work with the vendors that build the systems to focus on it, so that there are standard open adequate platforms for applications and users.

Let's talk about the applications and what people really want to do with these things. I'll talk about what we're putting into them, I'll tell you about what we're building, show you what we're building, talk about what some of the features are in the strategic period, and then tell you how we're building them so you get some appreciation for what we're trying to do and to see if you believe we can accomplish our task.

First is what we're putting in. We are focussed on three key platforms I mentioned earlier: the Macintosh, the PC using Windows and DOS, and the PC using OS/2 and Presentation Manager. That's not to say that there won't be others some day and that we won't support them. I often get asked the question about UNIX, so I'll answer that question. I don't have anything against UNIX, and I only have two requirements: One is that it has an adequate set of facilities, i.e. it has a graphical user interface, and a set of capabilities that allow the applications that operate in the mode that we've designed the applications. And secondly that it has a user base that is large enough to support my investment. In other words it's just a business decision. And right now, with the number of UNIX systems and the confusion over the user interface, and the lack of volumes in just absolute numbers, it's just not a viable business for me to be in. It's not a religious war, it's not a crusade, it's just a business decision. So some day if it changes we might look at that.

What we want to build in also is consistency across platforms. As you showed me by raising your hands this morning, there are lots of you who have mixed environments. And it's important for productivity tools and group activities at least, that the tools that people use can be used corporate-wide in this future environment. Because the expense account form that people fill out or the budgeting information that they do or the personnel records that they keep using on these systems need to be interchanged between people who have one or the other type of system. You shouldn't have to select the system based on the applications it runs. There are other characteristics of the system that you choose to select it on. And that's not only the user interface, it's also data storage and features and capabilities and macros and all other kinds of interfaces that the user would have with the system.

The third characteristic is that we want to be able to interchange and integrate data across the applications. This is a particularly important area because what's happening today is that applications are just getting larger and larger and larger. And you probably have people at your place who use a spreadsheet for their word processor. Or use a spreadsheet for their database product. And if you start looking at the requirements, you get the requirements lists for Excel, and on the requirements list is a spelling checker and rich formatted text. You look at the requirements list for word processing and it's tables, which by another name is rows and columns of numbers. And you look at the presentation graphics system, and not only do they want to make slides, they want to do text, they want to do spelling checking, they want to do charting, they want to do tables, they want to do drawing. And pretty soon we're going to have every application to do every feature. And that's not really an efficient way for me to develop and for your users to deal with these things.

In the long term we will get much more to a object tool metaphor where you have a set of tools that works on the thing you need to work on, but there will be a transition. And the transition will be the ability of the users to interlink and choose the facilities for multiple applications and integrate them. And we'll show you that. That's not just a trivial marketing concept, that's a full development concept of how we move from where we are today to an object-oriented environment.

We know that much of your data is not in your workstations, and so we have to provide easy access from the applications you want to do on the workstations, be that spreadsheets be that databases, whatever, to the data where it exists.

Output - I don't think there's such a thing as a paperless society. In fact, if anything, the quality of output has continued to grow with the capabilities and affordability of the I/O devices. I didn't ask, but how many people use laser printers primarily as their printing devices? It wasn't four years ago that no one had a laser printer. They used dot matrix and that was good enough in its day. And now it's laser printers, and soon it will be color, and it might be some other set of capabilities that become economically affordable. So we shouldn't forget that much of this information gets rerendered back into some form of output.

Ease of use is a key characteristic that I will spend a couple minutes on later. We're going to have all of these products be more programmable. Today if you did a survey, I would venture to say that in your organization probably less than five percent of the programming energy is done by the end users versus the professionals. I would venture to say that at the end of the 90's, you count up the amount of code that gets executed, that some much higher percentage, maybe even up to 80% of the programs that were written in the 90's will be written by individuals, not by programmers. And what needs to take place for that to happen is they don't want to become programmers, so they need to develop code or programs or routines as a by-product of their work, not as programmers. And we will be very successful if they write programs, and if asked the question do you know how to program, answer "no". So the objective we have is not to make everybody programmers but make the systems work much more like the people, and have the people be able to adjust the programs to their use.

And last, but not least, is an open environment. Even if you embrace my strategy and embrace the applications that I'm working on, it would be naive to believe that you're only going to do business with Microsoft. So we want to build a system that is open so that you can choose and pick the pieces that fit the business environment that you have, knowing that no vendor is going to have them all. And what that really means is, if you choose not to use my word processor or not to use my presentation graphics system, I'm still going to be open in providing the links for my products to link to and deal with other peoples' products. And that gives you choices, and allows me to keep a focus on competing on each piece. If my pieces aren't good and competitive in total then you shouldn't buy them just because I have the full suite, or because I have six or seven applications. You ought to look at every product I have and make a decision that that product is the best product on the market or as good as you need for your activities.

So let's talk about what we're building. We really have focused our development applications energy in six depth categories and one integrated/breadth category. We have four platforms that we build on. We have one character-based platform where we build OS/2 and DOS applications. And we have three graphical platforms that we build on. And we have those application areas.

Let me spend just a couple of seconds on those, and what I want to do is just get a little bit further into the future than the next release and talk about each of them and what they might do in a future time frame.

I think word processing probably will not continue to evolve towards desktop publishing. I think we've probably gone as far, right now at least, as many people will want to use in those kind of things. I think the next horizon for word processing is the assistance that most people need in the way that they construct and write - grammar checking, sentence structures, grading of the level of the readership or the level of how you wrote the document. I think that's something that a high percentage of the people can use, it's a very trivial thing to learn. Better spelling checkers and those kinds of things. You'll see a lot of features in the word processors we're going to show you this morning. I thing that we have kind of homed in on the set of features that most people use most of the time. And we'll find more and certainly there will be new features in every release, but I think the revolutions won't be in better typesetting or better fonts, but much more in other areas of processing.

Spreadsheets - the vast majority of the world will use two-dimensional spreadsheets. They'll use things like Excel. And they need to be easy to use and easy to understand. We live to a great extent in a two-dimensional world in terms of how we do things. And that's represented by paper, that's represented by blackboards, that's represented by slide screens. There is a percentage of the user base that needs multiple dimensionality. That may be in the 10-30% range. Now when I say two dimension, I include in two dimension the linking and consolidation of information. I think that's a two-dimension function. And I think consolidation is not just a dimension, it needs to be a much more intelligent function in the ability to take and consolidate data that's supposed to be consolidated and just not positional data. Not necessarily all the 'Row C3s' together, but all the rows that represent 'accounts receivable for 1989'.

Multiple dimensionality is an interesting topic. And you will hear a lot about dimensionality in the coming months from a lot of vendors, and what the benefits of it are and how they are going to use it. Let me tell you what the research that we have done in multiple-dimensionality indicates. The first thing it indicates is that three is an arbitrarily wrong number of dimensions. And you're probably going to chuckle when I tell you what we think is right, but let me give you a little background on why we think it's right. We went to some folks who have spent their life in the consulting industry and have dealt with what business planners and decision support people need. We've also had some discussions with some of the business schools on how they recommend how people use decision support and planning and where dimensions fit into that. It turns out that almost unanimously their recommendation is that the model needs to be five dimensions.

And you say, "Well, how does anyone get around five dimensions, and what are those five dimensions?" Let's take a business problem. You manufacture something. If you take a look at what you need to do when you get ready to do decision support or planning on your activities, first thing is that you have a set of products. Secondly, you probably have a marketing organization or some kind of characteristics like regions or divisions or major market segments. So you have some definition of market. You probably have some definition of where it's manufactured. And if you're doing planning you might make equal one area with some place else. So there's a dimension to your planning and what you have capability of planning.

You always have a dimension of time: periods of times, quarters, months, years, whatever. And often you'll have a dimension of cost. What does it cost, what are the financials involved. Now lets talk about how you might use that.

When you view data, you always view it in two dimensions. There aren't many of us here that have five-dimensional viewers in our mind. And what you think about is that you've got this chunk of data and you want to look at it 'products by region', 'sales by region'. Or maybe 'products by plant', or maybe 'products by time'. Or maybe 'costs by time' or 'plants by time'. And so it's taking this data that you've got and looking at it two dimensions at a time. If you take your planning document that you have at your corporation, you'll have sections. You'll have 'products by region'. You'll have 'products by month'. You may have 'regions by month'. And each one of those becomes a two-dimensional section in your control book plan.

And so I think that the two characteristics that you have to have is, one, the ability to view dimensions two at a time. Any two that you have in your database or in your spreadsheet data, two at a time. The second thing you have to have is be able to do goal seeking. And you want to keep all the data you have in your multiple dimensions consistent. Let's say for example that you want to increase your total sales from $90 million to $100 million next year, and you want to try to allocate that in your plan. So what you do is to look at some dimension and you say I want to have the answer total be 100, and now let's look across the regions. That's a $10 million growth, and let's put $6 million of it in Region A, and $4 million of it in Region B. The system needs to say, "What products are you going to sell more of?" So we'll go and look at the products. And let's say we decide we're going to sell half of product One and half of product Two. And the system says, "Wait a minute, do you realize that you don't sell product Two in Region A? How are you going to achieve that goal, are you going to open a new market?" And we say that we can't constrain it that way, I've got to change my products or I've got to change my regions, or I've got to change my time frame. It's all very interrelated. And when you do goal seeking in two dimensions or even three dimensions, you really lose track of all of the constraints you have in the other dimensions. So a spreadsheet that does multiple dimensionality has to allow you to manage goal seeking in all the dimensions and make sure that they all stay consistent and accurate.

In terms of presentation graphics, I think the important element is not the graphics, it's the presentation. And I think that what you have to do in presentation graphics is make it very easy to use in the construction of a presentation, not in making the presentation maximumly pretty. We'll figure out how to make it maximumly pretty also. But it's really important that you can construct and create presentations quickly and easily.

Database - there are two database things. One database is what is done at the system division, the SQL remote engine. What the applications area for database is, is the front end, or the user capabilities that deal with data. I think there are two or three things that happen there. One is that users write programs, or they keep small databases that they massage themselves. Secondly they buy programs that are supported by their group from their IS shop. And thirdly, they want to do queries and data analysis on data that they don't own, data that's stuck away on the system some place. And the important part of the database is graphical-based forms orientation, and nice reports and nice queries.

In terms of project management as a category, I expect a great deal of growth. We all do project management. We usually don't use project management tools because they're too complex and too project "managie". We usually use calendars or we usually use notebooks, or we usually use something else to plan the projects that we do. And what you'll find is that people put on their calendars, e.g. three days before a meeting, to write the agenda for the meeting. And so project management certainly has a requirement to deal with high-end projects, but also could be the application that a lot of people use more daily and is more central to their activities.

And the last is the mail system. We talked about what's important there a little bit when we talked about what groups do. Work flow is certainly important. Another thing that is important is that the system be able to deal with and manage forms. A great amount of stuff that flows around a system today is forms-based: expense accounts, time cards, phone messages, whatever. There are two main reasons that you have forms: One is that it kind of helps you figure out what you need to fill out, what's needed in the transaction. And just as importantly it helps you prioritize and bundle and characterize and categorize the things that you want to process. For example, you might not want to sign your expense accounts randomly. You might want to work on all the expense accounts that were submitted by your employees at one time.

As you move from ten messages a day to a hundred messages a day to several hundred messages a day in an electronic mail system, the problem changes in terms of your ability to deal with the information you're getting. So the mail system has to be much more clever in analyzing, categorizing, and bringing you a set of activities or a set of information that's on a single problem when you want to work on that problem.

The last application area we do is called the breadth application area. It is an integrated application that is built by combining pieces of all six other areas into a single product. That product is generally there for people to learn on, for people to use at home, for people to travel with, or for people to use at school. The product that we build in that category is Works. We don't believe that a depth application should be an integrated application. We don't believe that we should integrate a very sophisticated spreadsheet with a very sophisticated word processor. You ought to have the freedom to choose between multiple vendors for the depth product, and then have them work together as if they were integrated.

Now that's kind of an interesting overview, but let me spend a minute or two to say what we're going to build out of that.

In the character-based systems we are investing in four areas. The way we chose to invest in these areas is that these areas are where people have either systems installed that need to participate or we have applications already there that have a user base. So to the extent that people use Multiplan or Word we will continue to enhance those products and bring and add new features to those. The same is true with Works. Mail is an important one, because you will for a long time have older systems that only have character-based capabilities and need to be able to participate in your networking things.

In terms of the graphical based systems, it is our intention to target all seven of the categories for all three platforms. And that's an interesting problem because that's a lot of work. We have 11 of those 25 X's announced and available to you today. So the question you probably ought to be asking is, "You're probably out of your head if you think you're going to do 14 additional categories of products, and then have annual or 18-month releases of those products, so how do you plan to have that set of applications available?" We plan on doing that through a couple of techniques.

The first technique gets at development efficiency. It's the concept we call core engines. What it does is we've taken an application, let's use Excel as an example. We take out all the parts of Excel that work exactly the same on any platform. We call that the core and we write that one time. There's one person who writes recalculation for spreadsheets. There's another person who figures out the data structure for spreadsheets. All of that's written in high level language and can be very quickly ported to any platform. Then we take the piece of the product that is unique to the platform and that is the characteristics that make it unique to the user interface, or in some cases some of the platforms don't have all of the features and facilities of the other platforms, and we either have to simulate them or not have those as part of the product. That we call the 20%, and in Excel's case it happens to be about 18-1/2% unique to 81-1/2% shared.

Now what that allows us to do, is what used to take 300% of effort, we now can do for 140%. So we're really not writing 21 graphical applications, we're writing 7, 80-percenters, and 21, 20-percenters.

The second thing that we're doing, is that for every one of them we're insuring that there's dynamic data interchange. On the Apple that's a little more difficult because the operating system doesn't have interprocess communication. But in many cases we're trying to simulate that in the applications, and in some extent using MultiFinder, even across applications. But we want the applications to share information in a very transparent way to the user.

The third thing we talked a little bit about is programmability, and we want to make sure that all the applications have programmability. Not only that they have programmability, but just like consistency of user interface, you want to have consistency of programmability. So you'd like to have the same syntax, the same language, the same types of things that you can do in any macro language, be the same as any other product. That just means that when a person learns how to write a macro in Excel, then when they get a word processor or they get a database from Microsoft, they'll know the basic fundamentals of how to program in that. And maybe even know 100% of how to do that.

Last, but not least, is usability. It is a very important subject. We've changed what we believe is the learning model that people ought to use in learning products. And what you'll see happen over time in our products is the first learning experience will be a computer-based training course. The system will lead the user in how to use the system. Secondly there will be a contextual help system that has multiple degrees of help, with the third degree being back to the training course. So you can get a summary help, a detail help, and if you still can't figure out to do this, you go back and do the training course for that section. That says that the paper documents that come with the product change their character. They don't become "how to do" manuals, they become reference manuals for when you want to do something very unusual or very special or get very detailed information. And so we reorient those to have an encyclopedic kind of organization. It's kind of like you learn the product, you use the product, and when you go to learn something special or more different or with greater depth you go to the library to check out the World Book. And if you want to look at printers you look at P's. If you want to look at setup you look under S, etc.

The third thing we're adding to all the systems, and we'll show you a little bit about it, is the ability for the user to build the interface that fits either the application or the user. Kind of a softer software.

Last, but not least, is some tech work we're doing on how people learn, and how you will use some of the technologies that will be available. If you study what a large number of people learn and how they use the system, what you find is that what you learn the first three or four days you had the product is often all that people know for the entire time they have the product. They go to class and learn how to do a few things. And from then on they use that set of tools, and they do things often very difficultly using that same set of tools, but that's what they know.

There are some people who are curious and will find out other things, and they do it usually one or two ways. They walk down the hall and they stick their head in their buddy's office and their buddy is doing something and they say, "Huh, that's interesting. How'd you do that?" And so they have a lesson on how to do one other thing. The next thing that happens some times is that you know your product can do an importation of a graph, but you're just not sure, so you call Joe the local guru and say, "Joe, how do I make this work?" And you almost always learn those things a little bit at a time when you need to use them.

So with the capabilities we see on the horizon, we've started a project and we'll do it initially on the low-end for the home, but it's certainly applicable to business. It involves taking that set of information and modeling how the computer might help learn these products and grow the user base in their knowledge. Let me give you a scenerio. You're using a spreadsheet and you want to add up three columns of data. And you enter equal, sum, parenthesis, highlight the range you want to add up, close parenthesis, and you've created a formula to add up that column. Now the user moves to the second column, and they start going through exactly the same scenerio, creating the formula to add up the second column. And then they go to the third column, and they start doing that scenerio. And then a little face pops up on the screen, and it talks to the guy and it says, "say, Joe, do you know about the copy command? It would really help you do this when you want to do the same function from one sheet to another sheet. So if you have a few minutes I'll teach you how to use copy, and you can really save yourself some time."

So the system becomes, using artificial intelligence, using video, using sound, the mentor, the tutor that would sit behind you and observe what you're doing, and helping you do that better. Teaching you how to use the advanced features of the system when you need to use the advanced features of the system; not as an intellectual exercise. In the strategic period we will be doing that, and the products you buy from us will have that capability.

Now, let me just tie this all together, and talk about how we see the products we're building fitting those user requirements. Of course we talked a lot about, and we'll show you a lot about individual productivity. In terms of workgroup support we kind of outlined what those are. The two things that we think are important elements that we're working on in the applications division for workgroup support are: one, modifying all the personal productivity applications to have features that can be used by groups; and secondly, tying it together with a mail system. The mail system is probably a distributed processing system that includes a lot of these other activities.

And last, but not least, is the business operations, and we don't plan to be in that business. We're not going to sell you payroll or chemical analysis or vertical applications.

We look at that as the two things that we want to participate in: getting you tools to write programs with - compilers, object-oriented compilers and so forth; and allowing you to have a sophisticated set of capabilities using macros and so forth. And we'll show you some of these things so that advanced users can write their own programs.

What all that means is that the users can complete their task faster, and do it more efficiently, and have better communications. All that measures in productivity. And they can do it with less training and less support which measures in reduced cost. We think that that bodes well for us as an industry, and for us well as Microsoft. Some day we may slow down, but there's still great growth ahead. And I think we're uniquely positioned, because we really do have a vision of what the industry might be, and we really want to try to use our resources to work on all of the problems within that. We've built control programs, communications, database, and the applications. We're not trying to say that we're going to solve every problem that you have. We're certainly not going to attack your host problems and the communications to your host, and some of those things; but we are trying to make the personal computer the platform and the front-end to all the information processing that your company does.

ACKNOWLEDGMENTS

Thank you, God—Father, Son, and Holy Spirit. I love you. Thank you for loving me.

Thank you, Susan. You are my best friend and my love. Thank you for journeying with me on this rocket-ride life. You have made our home a stable place for our family and even beyond that for our children's friends. You are mom to so many! The rocket ride has been a rollercoaster at times. Thank you for loving me through this crazy ride! I will try to lessen the steep drops on the road ahead. I love you.

Thank you, Jennifer, Amanda, Sarah, and Jonathon. You all have brought such great joy to mom's life and mine. God blessed us with you. And you bless us in being who God created you to be. Thank you for your love and for loving great mates: David, Chris, and John. Thank you for Elijah, Cecilia, and Owen. A new chamber in our hearts opened because of them.

Thank you, Dad. You are my hero. You taught me so much about life and principles and continue to do so. I am blessed to be your son.

Thank you, Mom. You too are my hero. Your gentle love and joy made our home an amazing place to grow up. You and dad taught us by your example.

Thank you, Monsignor Stan. You have been another hero in my life. Ever since I can remember, I have admired you as a man of God and a person filled with love for life. I will always remember our motorcycle trip together to the Black Hills of South Dakota. There were times when you became a speck on the horizon as you hit the accelerator and enjoyed a high speed cruise with just you, God, the wind, and a long straight prairie road. I couldn't catch up to you then, and your life and example have also been like that for me.

Dad, Mom, and Monsignor Stan, you have led so many to a personal relationship with God. I admire you, your steadfast nature, your work ethic, and your passion for Jesus.

Thank you, Rich MacIntosh. You gave me an opportunity for which I am forever grateful. You mentored me. You led by example. You cared. You made a huge difference in my life.

Thank you, Bob O'Rear. You, too, gave me the incredible opportunity to be part of an amazing team. Your gentle leadership and friendship helped me in more ways than I can say.

Thank you, Bill Gates. Your vision changed the world. And your work through the Gates Foundation continues that legacy into world health, raising up women, and in many other ways making the world a better place. Thank you for the opportunities you gave me, for your intellect and vision, for your support and your friendship.

Thank you, Jon Shirley. Your business acumen is second to none. Your leadership helped all of us be better and reach higher.

Thank you, Frank Gaudette. You were called home too soon. I am thankful for you and all the ways you supported Susan and me in those early days, including the personal ways you showed you cared. You lived "people are our greatest asset" in all of your actions.

Thank you, Scott Oki. You taught all of us about strategy in ways that have affected everything we have done since. Thank you for the support you gave me through all our years of working together.

Thank you to every team member who I had the opportunity to work with over these amazing years. It was a team effort all the way. I am blessed to have had the privilege of working alongside you.

I want to extend special thanks to the team that helped make this book a reality:

First to Susan, thank you for your numerous reviews and suggestions and being willing to relive some of the joys and hurts from this time of our life so that others can hopefully benefit from our experience.

Thank you to "Big Win Leaders" Joe Simon and Jeff Lockert for "edit round two."

Thank you to Terry Whalin. Getting to know you over sixteen years ago was a tremendous blessing. Your leadership as a writer and a coach to writers is inspiring. Thank you for believing in this book. And thank you for referring me to Bill Watkins.

Bill Watkins, thank you for all of your efforts as editor to take a good book and make it great and understandable, especially to those who may not be techies.

To the team at Morgan James, thank you. Thank you for believing in this book and working to get it to as many people as possible.

Your efforts have created the opportunity for many more people to hear these stories and learn from the lessons of the journey.

Thank you, Steve Jobs. You too went home too soon. Your passion, vision, and competitiveness revolutionized our lives and made the computing devices that surround us truly personal. The impact you and Bill Gates made changed our world and will continue to impact the world for all future generations. Iron sharpens iron. Microsoft and Apple sharpened each other.

Congratulations to Microsoft, Apple, Google, Amazon, and other technology industry friends and alumni. It's a new and great day! Thank you for making the technology that has changed our world for the better and enabled new possibilities for our collective future. We live in amazing times.

About the *AUTHOR*

Dave has been married to his beautiful wife Susan for thirty-four years. They have four adult children, three sons-in-law, and three grandchildren.

Dave Jaworski has over thirty years of sales, technology, and executive management experience. He has a bachelor's degree in computer science, and has spent half of his adult life on the technology side of business and the other half on the sales, marketing, and operations side.

He was the third employee at Microsoft Canada, and served as the national sales manager before being promoted to general manager of Western US Sales. There he led the team managing 80 percent of Microsoft's US business. Dave was promoted to GM of US Sales Operations and then led a turnaround as head of Microsoft University. He received the first ever *Bill Gates Chairman's Award of Excellence.*

After leaving Microsoft, Dave served as VP of Sales for Arabesque Software where they developed, launched, and took the top position in Personal Information Management software with ECCO at the world's largest software distributor, Ingram Micro.

He founded Provident Ventures, Inc., and helped GTE (now Verizon) launch their business in new regions across the country, being nominated for their Chairman's Award—their highest honor reserved normally for employees.

After Provident, Dave became the senior vice president and general manager of Gaylord Digital, the Internet division of Gaylord Entertainment. While there, he and his staff launched the Dixie Chicks first ever headlining tour, ran the largest Christian music store on the Internet, did the first ever Internet broadcasts of the Grand Ole Opry, wired the historic Ryman Auditorium in Nashville for web broadcasting, ran the independent artist network Songs.com, and streamed World Youth Day from Rome, Italy, over Real Networks and Windows Media networks to the world.

In 2003, along with music industry executives and leaders from the Gaylord Digital team, Dave founded and became CEO of PassAlong Networks, which provided digital media and content management for over two hundred clients, including eBay. They also created patented and patent-pending technology for eCommerce and sharing music.

PassAlong was the first company in the world to innovate and implement micro-payments with PayPal.

From 2009 to 2011, Dave Jaworski served as chief technology officer for Intero Alliance, building the Intero Lifestyle Network, a content management system and digital media store platform with extensive integration of global social media. Intero delivered a customized implementation of this technology to Avon Products, Inc., the world's largest direct sales company. Intero Alliance operated this platform for Avon in sixty-two countries and thirty-seven languages.

From 2011 to 2013, Dave served as vice president of Sales and Marketing for NetSteps LLC, selling the ENCORE™ solution to direct sales companies. ENCORE™ offered Cloud-based, Software-as-a-Service, Hosted Solution, and Enterprise licenses. Clients included Natura, PartyLite, Beautycounter, Scentsy, Rodan + Fields, and Synergy Worldwide.

In 2013 and 2014, Dave served as a member of the Board of Directors for PNI Digital Media and helped the company achieve a 7X return for its shareholders in that period with their acquisition by Staples, Inc. Staples' executives said it was their most successful acquisition ever, with 100 percent employee and management retention and over 99 percent shareholder approval.

Currently, Dave Jaworski is the CEO of Meta Media Partners, serves on the Board of Directors for 2X Global and on the Advisory Board for Cure Violence.

M·J
Morgan James
Speakers Group

Printed in the USA
CPSIA information can be obtained
at www.ICGtesting.com
JSHW022328140824
68134JS00019B/1360

9 781683 504207